Calm in the Chaos

Calm in the Chaos

True Tales from Elite U.S. Navy Aviation Rescue Swimmers

Brian Dickinson

Essex, Connecticut

The views expressed in this publication are those of the author and do not necessarily reflect the official policy or position of the Department of Defense or the U.S. government.

The public release clearance of this publication by the Department of Defense does not imply Department of Defense endorsement or factual accuracy of the material.

An imprint of The Globe Pequot Publishing Group, Inc.
64 South Main Street
Essex, CT 06426
www.globepequot.com

Distributed by NATIONAL BOOK NETWORK

British Library Cataloguing in Publication Information available

Library of Congress Cataloging-in-Publication Data
Names: Dickinson, Brian, author.
Title: Calm in the chaos : true tales from elite U.S. Navy aviation rescue swimmers / Brian Dickinson.
Other titles: True tales from elite U.S. Navy aviation rescue swimmers
Description: Essex, Connecticut : Lyons Press, [2024] | Includes bibliographical references.
Identifiers: LCCN 2024028489 (print) | LCCN 2024028490 (ebook) | ISBN 9781493078530 (cloth) | ISBN 9781493078547 (epub)
Subjects: LCSH: United States. Navy—Search and rescue operations. | Rescue swimmers—United States—Biography. | Helicopters in search and rescue operations—United States. | United States. Navy. Helicopter Anti-Submarine Squadron 2. | Dickinson, Brian. | United States. Coast Guard—Search and rescue operations. | Special operations (Military science)
Classification: LCC U167.5.S32 D53 2024 (print) | LCC U167.5.S32 (ebook) | DDC 356/.160973—dc23/eng/20240705
LC record available at https://lccn.loc.gov/2024028489
LC ebook record available at https://lccn.loc.gov/2024028490

The paper used in this publication meets the minimum requirements of American National Standard for Information Sciences—Permanence of Paper for Printed Library Materials, ANSI/NISO Z39.48-1992.

To my brilliant wife, JoAnna, who provided
incredible contributions of wisdom!

and

To the very few who choose to risk it all to
save another. Hero is an understatement.

So Others May Live

Contents

Foreword

As the sun rose from the east over the Himalayan Mountains on May 15, 2011, my life was forever changed as I took my final steps to reach the roof of the world. At 29,035 feet, I stood on the summit of Mount Everest completely alone. At the time, according to the Himalayan database, I was one of only two people to have ever had the summit completely to myself on a single day. I took in the 360-degree views, trying to process the magnitude of the moment in the short amount of time I could safely survive in the death zone. The death zone is the area above 26,000 feet where the pressure of oxygen is insufficient to sustain human life. After making a radio call to climbing teams lower on the mountain and literally taking my final look around, I began my long, slow journey down. Then, only 10 yards into my descent, everything turned excruciatingly bright white. I was completely snow-blind!

I immediately dropped down and grabbed the fix line, which are ropes attached to anchor points in the rock and ice and assessed the situation. I was at the highest point in the world. I was completely alone. I was blind. And nobody was coming to save me. Without panicking, I stood up and started making my way down. Hand-over-hand I felt my way down from the highest mountain in the world. What should have taken three hours to descend to high camp would end up taking me seven hours. I took a major fall off the Hillary Step and the South Summit, but the rope I was attached to shock loaded and saved my life. After over thirty hours of climbing, I eventually ran out of supplemental oxygen. But through unrelenting determination, faith, focus, and a miracle, I survived the impossible.

I wouldn't regain my eyesight for over a month, and I lost over twenty pounds. Weeks later after I returned to the United States, my survival experience was picked up by the media and I wrote a book called *Blind Descent* detailing my story. I've since become a motivational speaker, sharing my story of survival with major companies, organizations, and schools. I explain how through faith and focus I was able to find a reason to take one more step forward to descend to safety. Throughout the years of speaking, I have recognized one particular area that I tend to gloss over in my talks, which many people have felt the urge to reach out and learn more about. That moment I realized I was blind and had to keep moving to get down the mountain should have caused me to panic and most certainly die. But in that moment of certain death, I was able to go back to my military training and compartmentalize that panicked response to keep focused and push forward.

In this book, I unpack the U.S. Navy Aviation Rescue Swimmer training and experiences that helped keep me alive. I describe how the Navy trains their most elite rescue swimmers to control panic and thrive in chaos to save lives. I share my personal experiences as well as never-before-told stories from past and present aviation rescue swimmers. However, this is by no means a complete representation of all who have earned the prestigious title. Everyone has a unique story. Mine and those I've interviewed are ones of a few thousand. If you feel anything I write is wrong or offensive, then I want to say from the depth of my heart that I don't care. Stop reading immediately and find another book. This one isn't for you. My goal isn't to tell interesting training and rescue stories for the sake of entertainment, but rather to abstract critical commonalities learned that may help others. These same commonalities can be taught and applied in our normal lives where we tend to panic in all sorts of situations. I'm also blessed to include my wife, JoAnna, as a contributor to share her perspectives as a Christian counselor. She and I have been together since 1995, and she's been through both my military and professional climbing experiences. Her wisdom is included throughout each chapter in italics, to provide an additional perspective on the aviation rescue swimmer experiences, which the average person can use to help overcome their personal fears.

So Others May Live—U.S. Navy Rescue Swimmer Motto

Chapter 1

Mindset

"Smoke located in the water at four o'clock for 1,000 yards! Come hard right!" instructs the helicopter crew chief, never losing eye contact with the victim in the water. Smoke or flares are used to signal for help in survival situations and also as markers from the helicopter crew.

"Copy that, coming hard right! Automatic approach checklist. Crew rig for rescue!" commands the MH-60S helicopter aircraft commander (HAC), as he increases the Sikorsky Seahawk helicopter's angle of bank and speed toward the victim in the water. They fly low by the survivor to mark their location by dropping a smoke, which they will also use to indicate the wind direction for their final flying approach to deploy the rescue swimmer.

"Survivor in sight. Looks to be face down in the water! Request an immediate 10 and 10!" The crew chief urgently informs both pilots and the rescue swimmer of the intensity of the situation. A "10 and 10" means the pilots fly the helicopter down to 10 feet and 10 knots, which is deemed a safe height and speed for a rescue swimmer to jump during a day operation. The alternative is 15 feet and 0 knots or lowering the swimmer down the wire from a 70-foot hover (height of hover is helicopter specific), depending on the wind, weather, time of day, and obstructions in the water.

"Roger that. Survivor in sight, commencing manual approach." The HAC responds as the rescue swimmer, dressed in a full wetsuit, harness, deflated flotation vest, mask, and fins, scoots to the open cabin and sits with his legs and fins dangling out of the door preparing to jump. The pilot reports the altitude and ground speed throughout the approach, "30 feet / 30 knots, 20 feet / 20 knots, 10 feet / 10 knots. Standby to deploy swimmer."

"Swimmer ready," communicates the crew chief as he grasps the back strap of the rescue swimmer's harness and, with an open palm, hits the swimmer's gunner's belt. This indicates to the swimmer that it's safe to release the safety belt that tethers him or her to a support ring in the helicopter.

"JUMP. JUMP. JUMP," directs the pilot as the crew chief promptly taps the swimmer's right shoulder three consecutive times. The swimmer looks left, looks right, and looks through his or her legs for any obstructions, then pushes off from the cabin door and enters the frigid ocean water below.

"Swimmer away." The crew chief waits for the swimmer to surface and give the swimmer OK sign. "Swimmer OK," he promptly informs the pilot as the rescue swimmer urgently swims through the high seas and hurricane force of the helicopter's rotor wash toward the drowning survivor.

"Roger that. Engage hover mode." The HAC moves the hovering aircraft up to 70 feet and back to provide space for the rescue swimmer to perform his or her safety checks and procedures with the survivor before signaling the crew chief for extraction.

* * *

Jumping out of a completely functional helicopter into the middle of the freezing ocean to save the life of someone we don't even know is just another day in the life of a U.S. Navy aviation rescue swimmer. You won't hear much about us in the media other than maybe a brief headline, lacking specific details of a military or civilian rescue. You won't find many, if any, books written about us. And most written or media references will mistake our identity with another military special operations team. You won't see us bragging about missions and experiences at the local bar. If you do, then it's most likely not one of us but an example of stolen valor. *Stolen valor* is the term used to describe military imposters or those who lie about their military service. In my experience, the ones doing all the talking are typically the ones who have done the least. Aviation rescue swimmers are a small, elite group of men and women who have chosen a life of sacrificing their own lives to save another. We are a close family. When we hear the news of a rescue, we celebrate the successes, but when we hear of a helicopter going down or another life sacrificed, we all feel it as we mourn our family's loss. Our motto, So Others May Live, is the ethos we live each day by.

* * *

A healthy mindset includes avoiding selfishness and greed as ambition and pride cloud judgment in your decisions, love, and unity with others. Instead, be encouraged to focus on the needs of others, which ultimately leads to a life of humility, yet still making space for your own needs and understanding of self. The rescue swimmer motto, So Others May Live, aligns with this focus by guiding us to walk our days in humility, which allows us to experience the love and unity required to serve others.

* * *

During the writing of this book, a Helicopter Sea Combat Squadron 8 (HSC-8) MH-60S helicopter crashed off the flight deck of the USS *Abraham Lincoln* off the coast of San Diego, California. The helicopter had a main rotor malfunction, causing extreme side-to-side vibrations, which resulted in the main rotors striking the flight deck. The helicopter then violently rolled off the side of the flight deck, crashing into the ocean 70 feet below. It hit with extreme force, filled with water, and then rapidly sank, killing five of the six crewmembers onboard. One rescue swimmer miraculously egressed from 25 feet below the water's surface as the helicopter sank to the bottom of the ocean with the five deceased crew inside. The heroes lost in the crash were pilot Lt. Bradley A. Foster, twenty-nine, of Oakhurst, California; pilot Lt. Paul R. Fridley, twenty-eight, of Annandale, Virginia; Aviation Rescue Swimmer AWS2 James P. Buriak, thirty-one, of Salem, Virginia; HM2 Sarah F. Burns, thirty-one, of Severna Park, Maryland; and HM3 Bailey J. Tucker, twenty-one, of St. Louis, Missouri. The aviation and rescue swimmer community were heartbroken. We were all writing on our private social media groups, texting, calling, sending prayers, and changing our profile pics to the rescue swimmer logo with a black line through it to show our support and to raise awareness. Our hearts were aching as we learned of more loss to our family doing what we were wired to do.

One of the lost heroes, AWS2 (posthumously promoted to AWS1) James "Jimmy" Buriak's friends and loved ones shared a past glimpse of the type of person James was, when he selflessly saved two tourists in Guam. On a typical warm, tropical day, he and a buddy were walking along Gun Beach in Tumon when they heard panicked cries for help coming from the water. Without hesitation, Buriak entered the water and ferociously swam to rescue the victims. One drowning person was caught in a strong rip current and was being pulled out to sea, but that didn't faze the highly trained rescue swimmer from endangering his own life to save another. James swam directly into the outgoing path of the rip current, which helped propel him quickly to meet with the victim. The completely exhausted survivor could barely keep his head above water. Buriak communicated his intentions and placed the victim in a cross-chest carry, providing much-needed relief for the tired survivor. He then towed him parallel to the shore to move out of the rip current and then turned toward the shore. Rip currents are strong, localized, and narrow currents, which move directly away from the shore. They are typically less than 80 feet wide but can be close to impossible to swim against. Most rip current–related deaths occur due to panicked swimmers fighting against the force of the current, becoming too exhausted to continue, and drowning. The best way to survive a rip current is to stay calm, yell for help, and if possible, swim parallel to shore to get out of the current and then toward shore.

By the time Buriak brought the final victim to shore, the local firefighters and Emergency Medical Services (EMS) arrived at the beach. They brought the exhausted victim to a triage area and assessed his medical condition. Later when asked about the rescue, Buriak's simple, humble response was, "I just happened to be the person there. I would think that regardless of who it was, they would've done the same." Buriak earned the Navy Commendation Medal for his actions of heroism to save a life while off duty.

I asked Jimmy's wife, Megan, to describe her husband: "Jimmy loved his family and lived everyday like it was his last. He held his people close and built memories that will last a lifetime. He was special; there was never a person he met that didn't like him. If you knew him, you loved him and that was Jimmy. He had honor, integrity, and courage. He fought when he was right and backed down when he was wrong, he worked hard and even if he didn't agree he would do what was asked of him. He was courageous, strong, and resilient. He was it, he always had it, whatever that took or was—he was your person. He was courageous both in life and in death. His devotion to his people—his family, friends, and fellow crewmen—has rarely been equaled. He is missed dearly, and the world is missing an incredible and stellar human being."

Megan launched the AWS1 James Buriak Foundation (https://theaws1jamesburiakfoundation.org) to provide pre-mishap education and post-mishap support to the Navy and Marine Corps and their families. The foundation also offers information, referrals, and serves as advocacy support for every Gold Star family. Gold Star families are those that lost a husband, wife, or adult child while they were serving in the U.S. military.

* * *

"Whether you think you can, or you think you can't—you're right."
—Henry Ford

In my experience, this quote is spot on. People either approach life assuming they can do something, or they can't. That perspective either inhibits their potential in life with self-imposed limits or can propel them to do great things even in the face of adversity. The psychology behind this is having a growth mindset or a fixed mindset. A fixed mindset is the belief that your intelligence and abilities are fixed and cannot be altered based on new knowledge, training, and experience, and a growth mindset is where you believe your intelligence and abilities can be changed through effort.

* * *

Every job in the military has its risks. The decision to volunteer and serve your country is very honorable in itself. Many military jobs put our service members in extremely dangerous situations, requiring intense training and experience to perform under pressure. Each position is unique, and I'm not taking anything away from the amazing heroes that serve at all levels of our military. My intention for this book is to provide you with the unique mindset, training, and techniques we hold as aviation rescue swimmers. I am including experiences from past and present U.S. Navy aircrew and aviation rescue swimmers, who continue to fulfill the military's specialized needs. The stories included are just a small sample of those that deserve to be shared, as we can all learn a lot about ourselves and how we address our own panic and anxiety by hearing the stories of others.

The other military branches have their own search and rescue resources specific to their mission-sets. The U.S. Coast Guard has aviation survival technician (AST) helicopter rescue swimmers. Their area of operation is typically around the coasts of the United States, but they can also be forward deployed. They must train for similar scenarios as the Navy but tend to rescue fewer downed military pilots due to their location and charter. The U.S. Air Force Pararescue service (PJs) perform both sea- and land-based rescue operations. They are trained both in combat search and rescue (CSAR) and as paramedics, and operate in all environments including parachute deployment. The U.S. Army and U.S. Marine Corps will also occasionally send a select few through Aviation Rescue Swimmer School training for specific mission requirements. The U.S. Navy and U.S. Coast Guard also have surface rescue swimmers, who attend separate training. They don't deploy from helicopters but instead from their respective naval or Coast Guard search and rescue (SAR) asset vessel, tethered to a line handler when entering the water. Surface swimmers are attached to ships and deploy from a Coast Guard cutter, rigid hull inflatable boat (RHIB), or j-bar davit on a Navy ship. A j-bar davit is an extended hoisting mechanism attached to the side of the ship. All the branches have unique specialization training with a lot of overlap. The special operators have mutual respect with one another, but also enjoy some healthy competition and bantering. All are elite and professional in what they do. And all are legit bad asses! Although we periodically train and conduct joint operations, the focus of this book is on Navy aviation rescue swimmers.

* * *

My wife, JoAnna, has practiced as a Christian counselor for over ten years. She earned her master of social work (MSW) with a dual focus on mental health and children, youth, and family from the University of Washington. She also earned her bachelor's degree in psychology with a minor in sociology from San Diego State University and is a member of the American Association of Christian Counselors (AACC). JoAnna's contributions, highlighted in italics, provide real-world insight, tools, and techniques that can be applied to your life as you navigate your personal obstacles. I encourage you to learn and apply what we have to offer, but to also educate yourself continually with what others in the military, first responders, and civilian professionals are willing to offer. We only have one life to live, so it makes sense to invest the time in understanding ourselves, acknowledging areas of growth, and continually learning to better our abilities and circumstances.

* * *

The U.S. Navy realized the need for a school to train helicopter aircrewmen in rescue techniques in the early 1960s. Each squadron conducted training to better equip their personnel to conduct rescue operations, but this decentralized method didn't last long due to the inconsistencies they were experiencing across the fleet. The training quickly consolidated into East and West Coast training facilities to better standardize the methods, equipment, and procedures. And then years later it was condensed to what remains today, the single chief of naval operations (CNO) approved Aviation Rescue Swimmer School at the Naval Air Station Pensacola, in Florida.

Initially, the aircrews were specifically trained for conducting rescues from the cabin of the aircraft as hoist operators, but not for deploying swimmers into the open ocean. The need for dedicated rescue swimmers was realized after the Vietnam War due to the great number of rescues required and alternative methods being utilized across the Navy. The first official rescue swimmer class was developed in 1971 in Quonset Point, Rhode Island. Prior to that, the Navy used other means to rescue and recover victims at sea. Even before helicopters were invented, flying boats or floatplanes would land on the ocean to extract stranded survivors. The PBY Catalina—a flying boat and amphibious aircraft used to recover downed airmen from the ocean—was introduced in the 1930s. The PBY Catalina is credited with saving the lives of hundreds of sailors and pilots from downed aircraft and sunk Navy vessels.

It wasn't until the 1940s that helicopters were introduced in the Navy for rescue operations due to their dynamic ability to fly in bad weather, hover, hoist survivors, and fly directly to emergency medical facilities. The early Sikorsky helicopters expanded their capabilities during World War II, the Korean War, and Vietnam War as the possibilities of new mission-sets were identified and tested. The Navy helicopters had dedicated aircrew, who performed the rescue duties by dropping life rafts and supplies, hoisting survivors to their cabin, or even landing directly on the water for extractions.

The first military air-sea rescue operation occurred in 1946 using a Sikorsky S-51. The new helicopter was in a demonstration for the Navy when it was diverted for an emergency to hoist a downed Navy pilot from the ocean. Two years later, HO3S rotary-winged aircraft were replacing floatplanes aboard forward deployed Navy ships to perform a utility role and plane guard for flight operations. After the outbreak of the Korean War in 1950, helicopters began transforming into combat operations to rescue personnel behind enemy lines. This is when combat search and rescue became an official military designation.

During the Korean conflict, Navy and Marine HO3S helicopters undertook the rescue of downed aircrew and the medical evacuation of injured ground troops. Among the pilots who flew the HO3S was Lieutenant Junior Grade John Koelsch. Despite poor flying conditions, dangerous mountain terrain, and enemy fire, he risked his life above and beyond the call of duty to attempt to rescue a downed Marine aviator in North Korea. However, his aircraft was shot down while in a hover, hoisting the injured pilot. Somehow, they both survived the crash and evaded capture for nine days in the dense Korean mountains. After being captured and imprisoned, Koelsch never gave up hope and refused to aid his captors. He died of malnutrition and dysentery a year later as a prisoner of war (POW). Four years later, Koelsch became the first helicopter pilot to receive the Medal of Honor for his heroism, awarded posthumously.

The first documented U.S. Navy helicopter swimmer deployment occurred on January 22, 1952, off the coast of North Korea. Aviation Machinist's Mate Second Class Ernie LaRue Crawford was serving as an aircrewman with Helicopter Utility Squadron One (HU-1). HU-1 was attached to the USS *Rochester* operating in the Hungnam area of Korea. Crawford voluntarily entered the frigid winter waters, well within range of enemy shore batteries and small arms fire. Due to the intensity of the crash and extreme exposure, the pilot in the water could not assist himself in the rescue. With little regard for his own safety, Crawford rescued the downed pilot in enemy territory. The helicopter aircrewman, teetering on hypothermia, worked diligently to cut the parachute loose from the downed pilot. Because of the poor equipment and techniques, Crawford's hands became too numb to continue cutting, so instead he signaled to the helicopter pilot to lower the hoist. The aircrewman attached the injured pilot to the rescue sling, who was then short-hauled to a nearby ship. Crawford remained in the freezing water for twenty more minutes, continually moving to stay as warm as possible. After dropping off the downed pilot, the HU-1 returned and was able to hoist the near frozen Crawford to safety and return him to the USS *Rochester* for medical attention. Crawford was later awarded the Navy Cross for his extraordinary heroism in saving the life of a downed pilot in the vicinity of enemy fire. His heroic actions also planted the seed for future mission-sets for helicopters in the U.S. Navy.

The next two decades focused on dramatic new aircraft enhancements, models, reliability, mission-sets, procedures, capabilities, and training. The military found new ways to use helicopters for search and rescue, vertical replenishment (VERTREP), aerial weapons support, medical evacuations, search and surveillance, anti-submarine warfare, and personnel transport. The first National Search and Rescue Plan and Manual was written in 1956, with accompanying training regimens to standardize procedures and techniques to increase efficiencies of SAR operations.

Established in 1967, Helicopter Combat Support Squadron 7 (HC-7) was tasked with a variety of combat, SAR, and logistical mission-sets. They give great credit to Paramedic Rescue Team One, NAS Cubi Point, Republic of Philippines, who was tasked with training helicopter aircrewmen forward deploying to Western Pacific hot zones. The aircrewmen attended Combat Aircrewman Rescue School, run by U.S. Navy SEAL Rocky Bliss. The intensive training included jungle environmental survival training (JEST), survival, evasion, resistance, and escape (SERE), combat swim school, aerial gunnery and weapons, advanced first aid and medical, hand-to-hand combat, and 10 feet and 10 knots helicopter rescue swimmer deployment. The freshly trained combat aircrewmen then deployed to support the conflict in Southeast Asia.

During the Vietnam War, Navy helicopter squadrons were attached to ships or operated as detachments on shore bases. Their primary mission was CSAR, which included both overwater and overland insertions and extractions, aerial weapons support, crash recoveries, and military personnel rescues. Navy helicopters saw a lot of action during the war, but they also suffered a lot of losses. Of the 11,845 U.S. helicopters deployed in Vietnam, 5,607 were lost to enemy fire or mishap. Numerous safety and tactical lessons were learned across the different military branches, which were then incorporated into the training regimens in the U.S. military's relentless pursuit of excellence.

Around this same era during the Apollo space exploration years, Navy helicopters utilized U.S. Navy underwater demolition teams (UDT) to recover astronauts returning from orbiting Earth and even the moon landings. Initially, the Navy did not plan to use UDT "frogmen" to participate in the actual recoveries. However, after a successful splashdown of America's second manned space flight, Mercury-Redstone 4, in July 1961, the Liberty Bell 7 spacecraft sank, and astronaut Gus Grissom nearly drowned when his space suit lost buoyancy from an open-air inlet. After that almost tragic recovery procedure, all ensuing Mercury, Gemini, and Apollo flights included UDT frogmen to attach a flotation collar and life rafts to the spacecraft before the hatch was opened.

On July 24, 1969, Apollo 11, containing Neil Armstrong, Buzz Aldrin, and Michael Collins, splashed down into the Pacific Ocean. The aircraft carrier USS *Hornet* was positioned 812 nautical miles southwest of Hawaii in preparation for the highly observed recovery. The Navy recovery teams were extensively trained and prepared for all contingencies, as the world was watching in real time. Helicopter Anti-Submarine Squadron 4 (HS-4) Black Knights launched their "old 66" SH-3D Sea King helicopter 66, which deployed the UDT swimmers into the water to extract the astronauts from their space capsule. Helicopter 66 is one of the most iconic helicopters in the U.S. Navy. The Navy later moved to three-digit designations, and Old 66 was redesignated Helicopter 740. However, due to its world fame, they would repaint it to Helicopter 66 for future Apollo recoveries. In her eight years of service, Helicopter 66 flew 3,245.2 flight hours. Sadly, on June 4, 1975, during a routine anti-submarine warfare training mission off the coast of San Diego, she crashed into the ocean. Her legacy remains today, settled amongst the deep, dark floor of the Pacific.

> *"The H-3 was a great vehicle for Astronaut recovery operations. It was dependable, with plenty of power, stable, had good range and a 24-hour capability. Just the sort of bird you'd want under you when you're carrying out an operation in front of a few hundred million viewers. We assigned 'Old 66,' as she became affectionately known, for all five of the initial lunar missions and she always performed flawlessly. For a short period of time, I reckon she was the best recognized helicopter in the world."*
>
> —*Chuck Smiley, CDR, USN, Recovery Pilot, Apollo 10 and 13*

After the UDT swimmers helped remove the precious cargo from the floating space module, the Navy aircrew would then hoist up the astronauts and safely return them to the aircraft carrier to be quarantined. The helicopter included a NASA flight surgeon who gave a medical evaluation on the flight to the ship. They landed the helicopter on an aft carrier elevator, which lowered to the hangar where the astronauts took ten steps on a red carpet to a mobile quarantine unit. The astronauts quarantined for twenty-one days in the early space exploration missions, since NASA wasn't sure if they might have been exposed to harmful elements from interacting with a foreign celestial body. Additionally, they wanted to protect any life they may have brought back from the moon in their lunar samples.

AW3 Michael Longe and his SH-3D crew sat on standby for the Apollo 10 and 11 missions. He also flew in the photography helicopter for Apollo 12, where they took pictures and video of the entire recovery. This footage can be easily found online with a quick search. For the NASA recovery mission, they had three HS-4 helicopters, with backup helicopters and crews on standby in case of mechanical issues, crew sickness, or other problems that could occur. The first two helicopters, Swim One and Swim Two, would deploy the UDT swimmers while Recovery One would hoist the crew for extraction and transport them to the carrier.

Apollo 13 was the seventh crewed mission in the Apollo space program and third that intended to land on the moon. Onboard Apollo 13 was Jim Lovell (Apollo 13 commander), Jack Swigert (command module pilot), and Fred Haise (lunar module pilot). The mission to land on the moon was aborted after a rupture and explosion occurred from oxygen tank number two in the service module. Fortunately, the mission controller crew back on Earth was able to improvise a solution to increase the odds of returning the crew. Due to excessive CO_2 in the cabin, they built a makeshift filter with extra parts in the capsule to reduce the risk of asphyxiation. For days they dealt with low power, freezing and wet temperatures, and minimal potable water as they circumnavigated the moon and steered a course back to Earth. With the severity of mishaps and infancy of space travel, Apollo 13's chance of survival was estimated at less than 10 percent.

Longe was the primary hoist operator in Recovery One to ensure all personnel made it safely to the helicopter to be transported to the carrier. Prior to the astronauts' return, the HS-4 aircrew spent months in Southern California training for the Apollo 13 Odyssey command module splashdown recovery. A splashdown was the primary method of landing a spacecraft, using three parachutes to slow its descent as it impacted the ocean. HS-4 worked with the Naval Special Warfare UDT swimmers and a mock space capsule in the Coronado Bay. They feverishly trained day after day to ensure they could perform the recovery in their sleep. They planned for every imaginable variable and meticulously worked within the strict boundaries NASA set for them. There wasn't just all the unknowns of space travel and protection of our space heroes, but the recovery would be in the media spotlight of the world during a critical time in the Cold War between the United States and the Soviet Union. There was absolutely no room for error, and the U.S. Navy professionals did everything in their control to ensure the United States would shine in the public spotlight. There were an estimated five hundred million people watching the recovery live on television around the world as NASA struggled to return the crewsafely.

The idea was to deploy HS-4 in the general area of the planned splashdown, assigned to a U.S. naval warship, and launch as soon as NASA alerted them of their contact, status, and location. Longe remembers the excitement and anxiety for the aircrew in the days leading up to the recovery. As the HS-4 aircrew anxiously sat ready on alert for the launch for recovery, they continually practiced in their minds the on- and off-script methods they would perform. Years later, Longe recalls the exciting moment and how amped with adrenaline they all were when they got the call that Apollo 13 was moving through reentry into Earth's atmosphere. It was go time!

The HS-4 Black Knights launched Swim One, Swim Two, and Recovery One from the USS *Iwo Jima*, an amphibious assault ship carrying helicopters and Navy and Marine Corps personnel. They were in constant communication with NASA Headquarters in Houston and the *Iwo Jima* as they anticipated the command module splashdown. They flew to the NASA-provided coordinates and established a circular flight pattern, waiting for news of the returning vessel. There is a four-minute delay in communications with returning space capsules because of the ionization of the air around the module on reentry. Due to the shallow reentry path required because of its damage, Apollo 13 increased the radio delay to six minutes—the longest six minutes everyone had ever experienced. If the command module's heat shield failed, they would have burned up like most meteors do before reaching Earth's surface. As people were starting to lose hope, Odyssey regained radio contact and HS-4 made visual contact with the descending vessel with three huge red-and-white-striped parachutes deployed. The world erupted in relief and joy!

Swim One and Swim Two flew close to the module and used grappling hooks to snag the parachutes. A grappling hook or grapnel is a device with multiple hooks attached to a rope. They are tossed at an object and dredged to either retrieve or tow it to a specific location. The naval aircrewmen worked diligently to detach and retrieve the three main parachutes to ensure they didn't create a hazard or mishap for the recovery. A parachute can quickly fill with water and sink, creating a lethal sea anchor for a watercraft or survivor. Once they had the three chutes away from the module, they were clear to deploy the UDT swimmers.

Swim One approached the floating space capsule, bobbing in the Pacific, and dropped to 10 feet and 10 knots. The crew chief shouted to the UDT swimmers to prepare for deployment as they sat in the cabin door eager to jump. One at a time, he gave the UDT swimmers three consecutive taps on the shoulder, and they then jumped from the forward-moving helicopter into the ocean. As they surfaced, they gave the OK hand signal and turned to swim toward the capsule. The first swimmer attached a sea anchor to the capsule to keep it from drifting, while two other swimmers attached a flotation collar to keep the unit steady and above water. Once everything was strapped, inflated, and secure, they deployed an egress raft for the astronauts to board.

The lead UDT swimmer took control of the raft and secured it to the floating module. One of the helicopters lowered the rescue net with the three astronauts' flotation suits. The lead UDT swimmer then opened the hatch and welcomed the space travelers back to Earth! He handed them the flotation suits for them to don in the module and then one by one helped them to the raft where they would be individually hoisted to Recovery One.

Recovery One came in for extraction and held a 40-foot hover over the egress raft. Longe lowered the rescue net, and the lead UDT swimmer assisted the first astronaut in and gave a thumbs up to Longe. He hoisted him to the helicopter cabin, ensured he was secure, and then hoisted up the other two astronauts. The swimmers remained in the water to assist with the module recovery by the ship. Recovery One made a wide turn back to the *Iwo Jima* to provide enough time for the astronauts to get a quick medical assessment from the flight surgeon as well as change into their blue NASA flight suits. Longe recalls the actual recovery being flawless with nothing out of the ordinary. They trained hard and channeled their adrenaline to focus on the mission, remaining calm and professional.

During this time, the demand for the helicopter crews was growing as the Navy began to conduct a wider variety of missions. The individual rescue training units agreed that it was time for the Navy to have professional rescuemen, but they had to justify the new designation by collecting and compiling information and material to support the need for such a rate (new position title and pay grade). The CNO had looked at the option several times but felt that such a rate wasn't warranted because of the limited community, no career path, and a lack of rotational billets. Despite the challenges, in 1970 the CNO approved a formal SAR crewman syllabus conducted on a limited basis. Successful completion of the syllabus results in a SAR specialty designation for the trainee.

Without official designation, SAR training began with HC-5 (West Coast) in NAAS Ream Field, Imperial Beach, California, and HS-1 (East Coast) in NAS Quonset Point, Rhode Island. HC-5 devised a syllabus approved by the CNO for training rescue crewmen. The training began with highly motivated enlisted men volunteering for flying and successfully completing a flight physical. They had to demonstrate their confidence and abilities as a first-class swimmer as defined in the Bureau of Naval Personnel (BUPERS) Manual. The rescue crewman syllabus took fourteen weeks to complete. The syllabus consisted of three weeks of swimming and first aid, two weeks of hand-to-hand combat, one week of Survival Evasion Resistance Escape (SERE) training, one week of night vision indoctrination and combat pistol instruction, three weeks of plane captain and rescue aircrewman duties, and four weeks of flight training.

The training continuously evolved through the next decade as lessons were learned. Originally run by the help of Navy SEALs, seasoned SAR crewmen took over the training and syllabus to fine-tune it to specific skillsets, techniques, disciplines, and equipment for mission-sets they would be responding to. In 1979, the CNO formally established the Search and Rescue Model Manager (SARMM). HC-16 assumed the duties for SARMM, based in Pensacola, Florida. The following year the Navy developed the first comprehensive Search and Rescue Manual (NWP 37–1) and Rescue Tactical Airborne Information Document (TACAID). Then, in 1983, Aviation Rescue Swimmer Schools consolidated from East and West Coast schools to a single location at Naval Air Station Pensacola. The program has considerably evolved throughout the years and is constantly in a state of research and development to ensure the rescue swimmers are appropriately trained for the future of naval warfare. The official and non-official names have changed from the early SAR wet crewman, air sea rescue, helicopter rescue swimmer, and air rescue swimmer, to the current aviation rescue swimmer. By the time this book is published, it may be called something completely different.

In 1984, the U.S. Coast Guard put a similar program together for their Helicopter Rescue Swimmer training. The Coast Guard attended the grueling Navy Aviation Rescue Swimmer School until 1997. During those years of joint training, they were building their own program to be more specific to their coastal needs, rescue types, and aircraft. The conditioning and techniques were similar, but they required more dedicated focus as their program grew and became more accepted and funded in the Coast Guard. Aviation Survivalman (ASM1) Steve Lurati was one of my instructors in 1993. He was tough but fair, especially when I pissed him off during a night rescue training evolution when I poked him in the eye during a head sweep before helicopter extraction. The Coast Guard training eventually moved to the Aviation Technical Training Center (ATTC) in Elizabeth City, North Carolina. (Coast Guard Rescue Swimmer training is currently being held in Training Center Petaluma, California, while the Elizabeth City ATTC is being renovated.) The helicopter rescue swimmer candidates are collocated with the Coast Guard pilot candidates for familiarization with the aircraft they'll operate in. Even though the training is now separate, the Navy and Coast Guard still perform joint fleet training, advanced SAR training, and conduct real-time operations together.

I served six years in the U.S. Navy as an aviation rescue swimmer (class 9319) with other duties like anti-submarine warfare, CSAR, and aerial gunner. Stationed with Helicopter Anti-Submarine Squadron 2 (HS-2) Golden Falcons (now redesignated Helicopter Sea Combat Squadron 12 (HSC-12)), I deployed on the aircraft carrier USS *Constellation* for two deployments in the Persian Gulf as a part of Operation Southern Watch. During my time in service, our squadron took part in several rescues, medevacs, special operations warfare, and countless carrier flight missions.

Aviation rescue swimmers are attached to helicopter squadrons, which deploy on either aircraft carriers or surface force ships. Helicopters are the first aircraft to leave the deck during flight operations and the last to land. They must be in the air and ready to respond to an incident at any given moment. This can range from a pilot ejection, civilian rescue, man overboard, to investigating unknown surface or subsurface vessels. Even when flight operations aren't active, the crews wait in Alert status, ready to launch within thirty minutes if there's a man overboard or some other unexpected incident.

Each flight, the crew put their lives at risk for the sole purpose of supporting and saving our military brothers and sisters. With the strenuous training, technical rescues in harsh working conditions, operating in all weather and war-related environments, even downrange under fire, the aviation rescue swimmer position is rated as one of the most dangerous jobs the military has to offer. They are relentlessly trained to remain calm in extreme situations to successfully complete the mission, while minimizing mishaps and casualties.

With so many military job options, what type of individual volunteers to regularly put their life at risk to save a complete stranger? Everyone is wired differently, and even those who feel they have what it takes quickly find out it may not be the best fit. Aviation Rescue Swimmer School holds some of the top attrition numbers in all the military due to the intense physical, mental, and academic requirements. Our main operating environment includes deploying in the open ocean and dealing with combatant victims. Most people don't work well in those conditions, and once put to the test, students have the option of either Drop on Request (DoR) or two attempts to pass. This means you can fail an assessment once, but the second failure typically results in an academic evaluation and rollback to another training class. Any failures in the second class results in disqualification from the course and being reassigned to another rating (occupational field) to fill fleet needs—that is, possibly being placed into a Navy job that doesn't have a rating or specialty, such as scraping barnacles or cleaning toilets. This constant fear of failure helps drive some to success, but it more commonly leads them to quitting and pursuing safer military jobs.

* * *

In the above example, those who gave up in training sometimes did so due to their fear of failure. They were so motivated to avoid failure that they chose to quit. Alternatively, there are those who persevered through their fears, even if they didn't succeed in the training. The outcome can be a sense of "I failed but that won't stop me from trying" (growth mindset) to "I failed and therefore I am a failure" (fixed mindset with shame). The key to mindset is perspective. How are you going to approach a setback? You choose growth over fixed mindset by facing your fear and working through that fear. What if the worst happens? What then? How can I go on? What can I learn from this experience?

* * *

During the intense, oxygen-deprived training, the candidates are put into constant danger, which pushes their ability to calmly handle complete, unpredictable chaos. Safety is a critical component of the program to ensure things don't get out of hand, and students have the option to quit at any time. In 1988, this gained a lot of media attention and a rework of the safety protocols when nineteen-year-old seaman recruit Lee Mirecki drowned during training. He was participating in a high-intensity rescue training exercise when things got out of control with the instructors ignoring his pleas to quit. Mirecki died from heart arrhythmia, ventricular fibrillation, and hypoxia. The school was temporarily shut down for investigation, and the safety protocols were revamped to prevent such an incident from happening again. And unfortunately, during the writing of this book, Airman Nathan Burke became the second fatality to occur in Aviation Rescue Swimmer training. Burke became unresponsive during a high-risk training evolution in the pool. No further details of the incident or investigation are available yet. It's a stark reminder of the dangers involved in the job. Unfortunately, all aviation rescue swimmers have experience with fallen brothers and sisters they've served with due to training, mission, or aviation mishaps.

The unfortunate training incidents certainly aren't the norm, but they demonstrate how close to death the candidates are at any given moment. Several students in my class blacked out in the pool or sucked water during a rescue, requiring training timeouts and in a few cases, resuscitation. One of my classmates was hit hard by an active survivor during a final blindfolded life-saving exam. He had the wind knocked out of him, he swallowed water, and then passed out. The safety instructor in the water pulled him to the side of the pool and revived him with rescue breathing, plus a few hand compressions to the stomach. Upon revival, an ambulance was onsite to ensure he was stable. He was then removed from the program.

The ultimate goal of the school's extreme training regimen is to weed out those who won't be able to fulfill the duties of an open ocean rescue, no matter the situation or conditions. The training is very physical, but it's also extremely mental. Physically you can usually work hard to overcome obstacles and build to the necessary fitness level. But if you lack mental toughness and can't remain calm to think through an intense rescue, then you will put yourself and others at risk. The instructors push the candidates well beyond their perceived limits to ensure they understand what they are truly capable of. Most people in this world live in a safe bubble, not knowing what they are truly capable of. Aviation Rescue Swimmer School ensures that the candidates are comfortable living uncomfortably. It all boils down to attitude: calculating risks, having a determined "no quit" attitude, and being confident enough to persevere in any situation.

Michael Bulman was one of my aviation rescue swimmer instructors in 1993. He was very tough and thorough, but he also had a positive impact on me due to his deliberate and sincere approach as an instructor. I recently asked him about his perspective on training future swimmers to handle panic.

"I think my focus was to learn how to take that panic and turn it into a positive input. I know most students didn't realize it but most of the head games we played with y'all were a tool used by many who came before us to help focus your attention on what you were looking at as a swimmer when you got the call to a rescue. It was a fine line that was very stressful not only for you as a student but for us as instructors. Anyone who says they don't panic when in that situation, doesn't have a pulse. As an aircrewman / rescue swimmer, you are in some kind of controlled chaos every time you step foot into an aircraft. We had to take all that, plus the curriculum of the rescue swimmer program and teach that to people who were not far removed from their senior year in high school. It was a most rewarding job to do. I absolutely loved everything about being in naval aviation, and I considered it an honor to have been chosen to be an instructor. But as for my pet peeve, it was attention to detail. Every single procedure we taught was written by our brothers who came before us, and they learned with their lives in some cases. I always took a little extra time on the steps that would kill you or the survivor if you missed them."

* * *

You must retrain your thinking to always move forward. Focus on what you can control, not what you can't. It's easy to wrongly focus on trying to control the things you can't control and then getting stuck in hopelessness or anger or sadness. The areas you can control are your thoughts and actions. You can respond accordingly to things that don't go right. Take the perspective of "How can I learn from this experience? What can I do to move forward?" If unsure, then seek guidance from those trusted around you. God created us for relationships because in relationships we learn the most about ourselves and others. Use the opportunity of relationships to grow yourself and how you respond to others and situations outside of your control.

* * *

A fixed mindset is the belief that your intelligence is fixed and cannot be altered based on new knowledge, training, and experience. A growth mindset is the opposite, where you believe your intelligence and abilities can be changed. Those candidates that had a fixed mindset didn't make it far in the training, since the skills learned are not skills most possess by default. The instructors constantly push the candidates far beyond their comfort zones, but with an open mind and the ability to evolve their thinking, they became experts in the field of rescue and survival. You can certainly survive this world living in a self-inflicted protective bubble, but you don't truly live until you push beyond your limits. It's at that point that your eyes open for the first time to see what you're truly capable of achieving. It's never too late to unlock your potential and reinvent the life you should be living. It just takes the right attitude, a growth mindset, and a little effort.

The few aviation rescue swimmer candidates possessing that attitude learned two key lessons listed on every technical training procedure. Two simple lessons tattooed in our soul, which made us the best of the best at handling fear and panic in extremely chaotic scenarios: Control your breathing. Don't panic!

Chapter 2

Face Your Fears

On a cold, moonless night in November of 2007, U.S. Navy Helicopter Anti-Submarine Squadron 2 (HS-2) Golden Falcons was conducting CSAR training in the mountains near El Centro, California. Due to its desert and mountains that mirror Middle East topologies, the Salton Sea just north of El Centro is a regular training ground for the U.S. military. In the 1940s, the base was used during World War II as a Marine Corps air station for training new squadrons preparing to deploy and for those returning to continue their training as they prepared to redeploy to war. The base was later redesignated the Naval Air Facility El Centro, which has two operating runways. Simulated aircraft carrier takeoffs and landings are conducted there as well as ordnance drops on the numerous target ranges. Helicopter squadrons conduct CSAR training and weapons delivery, plus joint military training operations between different branches and allies.

Two HS-2 HH-60H helicopters carrying seven aircrew each completed their nighttime training operation in the mountains east of the Salton Sea. The HH-60H Sikorsky Seahawk helicopter is a mission-specific model of aircraft used for U.S. Navy special operations missions. Its cabin is gutted of any non–special operations equipment, to maximize space for passengers, insertion and extraction gear, and a wide range of tactical weapons. The helicopter is equipped with forward-looking infrared (FLIR) cameras to allow the pilots to see through darkness, haze, smoke, and poor weather, plus for targeting Hellfire missiles. The HH-60H was replaced by the MH-60S in 2002, an aircraft with upgraded avionics, weapons system, and cabin sliding doors on each side. The Navy is currently researching the next generation of helicopter to replace the MH-60S and MH-60R.

The two CSAR helicopters flew in formation at a thousand feet due west toward the Pacific coast. They were low on fuel and radioed the Naval Air Station North Island tower to request permission to land and refuel before heading back to the USS *Abraham Lincoln*, which was operating 100 miles west off the coast of San Diego. Both aircraft landed and taxied to the fuel pumps where they hot pumped, meaning they refueled with the engines and rotors turning. The crew chiefs of both helicopters jumped out while on the deck and did a walkaround to check for any possible visual damage or concerns. None were identified or reported. Both aircrews then conducted their takeoff checklists and departed the base to head back to the aircraft carrier.

The helicopter internal conversations were relatively quiet during the return flight, since all were exhausted, hungry, and looking forward to grabbing a late dinner from the ship's forward galley before getting some rest. Wearing night vision goggles (NVG) and peering out the cabin window, aviation rescue swimmer Jesse Hubble's view of the peaceful waves of the ocean was suddenly disrupted. A massive green explosion of sparks burst from the other helicopter's tail rotor, green because NVGs turn everything green as the image intensification screen is made of phosphor, which glows green when struck by electrons. Immediately, "AUTO! AUTO! AUTO!" echoed through the radio from the helicopter aircraft commander flying the crashing helicopter as they autorotated down to the high seas below. An autorotation is where the pilot drops the collective control, which controls the pitch angle of the main rotor blades to increase or decrease power. This forces the helicopter to drop straight down in an attempt to control the crash, using gravity and the force of wind to turn the rotor blades rather than using engine power.

The HS-2 CSAR helicopter forcefully crashed into the water, slamming the seven crew inside the cabin with tremendous g-force. The impact of the crash automatically triggered the emergency locator transmitter (ELT), sending alerts over 121.5 MHz and 243.0 MHz frequencies to all monitoring radios that a major mishap occurred. The helicopter smashed down on the water and momentarily settled before sinking and filling the cabin with 63-degree ocean water. The seven aircrew remained strapped in as the aircraft filled with freezing water, rising up their legs and to their chest. The engine sounds were deafening with the free-spinning rotors growing closer to the water. With the bobbing of the waves, the main rotors began to tilt off balance and then violently struck the ocean's surface. The brutal force of the rotors chopping into the sea ripped them off, firing debris across the moonless horizon. The top-heavy helicopter then tilted to the port side, completely flipped over, and sank upside down into the black abyss. Each highly trained aircrew inside the sinking coffin egressed out of their primary or secondary exit, some with a couple breaths of their HEEDs (helicopter emergency egress device) spare oxygen bottle. Once on the surface, they activated their survival vest flotation by pulling the CO_2 activation cord and affixed strobe lights to their helmets. Everyone in shock from the unexpected crash swam to form a group a distance from any possible wreckage. They communicated their condition with one another and then one of the rescue swimmers ignited a MK-124 MOD 0 signal flare to help the other helicopter locate them in the pitch dark.

In the airborne HH-60H, Hubble automatically shifted his mindset to the unexpected mission of rescuing his teammates. With no prompting from the pilots or crew chief, the rescue swimmer donned his rescue gear and prepared for a nighttime water entry. This included wearing his long wetsuit, harness, mask, and fins, plus breaking a chemical light to place in a slot on the top of his mask. The pilots made a standard SAR approach for a nighttime rescue deployment. They flew the helicopter over the seven survivors' position and the crew chief opened the cabin door and tossed a MK-58 MOD 1 smoke/flare to mark the position and check the wind direction and sea currents. The pilots then flew a wind line rescue pattern to approach with the helicopter facing into the wind. They held a 70-foot hover to deploy the rescue swimmer. Once the swimmer was in the water, the pilots would then move the hovering helicopter left and aft to maintain visual contact from a distance, ready to move in for extraction of the personnel in the water.

Once in a stable hover, the pilots transferred flight authority to the crew chief's hover trim control, which is basically a joystick with a pressure-activated thumb control to move the hovering helicopter. From the open cabin door, the crew chief has the best view of the situation in the water. He can move the HH-60H closer to or away from the rescue swimmer to optimize the hoisting process, while communicating to the pilots his every move. Hubble sat in the cabin door, as the crew chief lowered the rescue hook to him. He connected his rescue harness–lifting V-ring to the large hook, and the crew chief lifted him up and out of the helicopter. He dangled there for a few seconds as he assessed the situation and then was lowered down 70 feet to the seven survivors huddled together in the frigid water. He disconnected from the rescue hook and swam toward the group.

Hubble established communication with the survivors over the loud HH-60H engine noise and heavy rotor wash. Fortunately, besides shock, there were minimal injuries to the aircrew. With a couple of his rescue swimmer peers in the water, they were able to work together to perform safety checks. Hubble maintained positional authority as he prioritized hoisting the pilots and passengers two at a time. After forty-five minutes in the open ocean, Hubble was hoisted up with the final survivor. The HH-60H cabin was packed and heavy with fourteen aircrew onboard. A flight medic completed medical checks on the crash victims as the HS-2 helicopter promptly returned to the aircraft carrier.

With each flight comes the possibility of an unexpected secondary mission. The aircrew and passengers were briefed for a normal CSAR training exercise in the Southern California mountains but ended up ditching the helicopter in the ocean due to a tail rotor failure. Every time the helicopter squadrons brief for a flight, they always include a section for SAR since it is always a possibility. When the horrific event occurred, Hubble didn't panic or lose focus even though he was concerned for his teammates egressing from the sinking helicopter. In life we can easily grow complacent in our daily tasks, but how we respond and refocus when things don't go as planned is what keeps us calm in chaos.

"When I am up in the sky, I own the back of the helicopter," Hubble said. "We train every day, physically and mentally, for any situation that might happen. We prepare for the worst-case scenario, and that night was one of the worst. It all happened so fast. You just do what you have to do. I am glad everyone made it out alive."

Aviation Warfare Systems Operator 1st Class Hubble was awarded the highest non-combat decoration, the Navy and Marine Corps Medal, for his courage in rescuing seven lives that night. "Relying on his special training, stamina and courage, he remained in 63-degree water for 45 minutes until all seven survivors were safely hoisted aboard the rescue helicopter. By his courageous and prompt actions in the face of great personal risk, Hubble prevented the loss of seven lives," the award citation read. Sadly, Jesse passed away suddenly a year after our interview. It was an honor to get to know him and include his amazing story in this book. Rest easy, hero!

Current training to become a Navy aircrew / aviation rescue swimmer lasts for almost two rigorous years. Before an official designation existed, the job was all on-the-job training by helicopter aircrew developing procedures as they went along. After the vital need for SAR swimmers and CSAR was realized in the Vietnam War, the rescue swimmer school of "hard knocks" created a specialized regimen based on their high standard of operations. Over the past decades, this training has significantly evolved and continues to evolve based on new techniques, aircraft, equipment, safety, lessons learned, and global response needs.

The long road to success can be daunting as you look at the relentless physical, mental, and emotional commitment. The Navy is investing a lot of time and money to ensure they have the right person for the job. They won't hesitate to drop a student from any of the courses if they feel they aren't exceeding the minimum standards or if they lack the commitment or attitude. Also, it's a volunteer position. Nobody forces or even asks Navy personnel to attempt the program. Each student is there by their own free will and can DoR at any time.

The time commitment and magnitude of training is extremely overwhelming, but breaking each phase down into small achievable blocks helps provide a more manageable approach. There are always setbacks, but it's how you respond to those setbacks that makes the difference. If you get frustrated and quit on the little things, then you'll never be able to hack it in the fleet. Training is beyond tough, but it's controlled and doesn't compare to the insanity that can occur later in the fleet deployed in a war zone. That's why its intensity only increases with each day as the trainees build up their fitness, rescue skills, and water confidence. The instructors often remind the students during pool training that there aren't any sides of the pool to grab onto in the middle of the ocean. So, you'd better not get caught trying to grab a side to rest, as the entire class will be punished. Rest only comes in the form of treading water, which becomes easier the more you do it. And you do it a lot!

For most special operations candidates, training begins during boot camp or Recruit Training Command (RTC) in a program called Dive Motivator (DIVEMO). Other candidates coming from the fleet must submit an application to Navy Personnel Command (NAVPERS) and be accepted after a thorough evaluation. Criteria includes a current flight physical, second-class swim qualification, and a physical screening test (PST) for Aviation Rescue Swimmer School (ARSS). Their last two Navy evaluations must be at least a 3.0 and recommended for retention or promotable, and they must have acceptable Armed Service Vocational Aptitude Battery (ASVAB) scores, Secret Security clearance, and no non-judicial punishment in the past twenty-four months.

As an effort to continually improve the programs, DIVEMO was recently replaced with an updated Aviation Rescue Swimmer School Preparatory Course (ARSPC). All prep for Basic Underwater Demolition / SEAL (BUD/S) training is now conducted in Coronado, California. I went the former route in my attempt at ARSS, at least guaranteed a shot after RTC, assuming I passed the prerequisites and PST. For us young recruits in RTC, typically fresh out of high school, we started our days a few hours earlier than the rest. At 4:00 a.m. each morning, while the rest of the Navy boot camp recruits remained nestled in their bunks, a group of early risers hoping to be aviation rescue swimmers (AIRR), explosive ordnance disposal (EOD) specialists, Navy divers (ND), special warfare combat-craft crewmen (SWCC), or Sea Air and Land (SEALs) gathered in the combat pool for early morning torture. Most of the naive candidates don't even come close to completing their respective training. In fact, many don't even make it past the RTC Dive Motivator training.

It's a good way to weed out the individuals early before the Navy invests further in the training.

During the dark, cold mornings the candidates line up in arm-length rows, hidden behind clouds of steam emitting from their heavy breathing. They just completed what felt like a million push-ups, sit-ups, flutter kicks, and pull-ups. The intimidatingly ripped SEAL and SAR instructors then lead the group for a two-mile run with frequent pitstops for more punishing calisthenics. The Dive Motivator instructors are current Navy SEAL, AIRR, EOD, SWCC, or divers assigned to RTC to educate recruits during boot camp and conduct special screening tests. The goal is to provide a more structured training program to better prepare the candidates for their respective special operations training.

After a grueling wakeup call on dry land, the candidates are ordered into the pool. As a team, they are properly conditioned with countless laps, both above and underwater. They tread water for ten-minute intervals while holding a brick above their heads and alternating by holding the brick between their feet for another ten minutes. They learn and perfect swim strokes for the sidestroke, breaststroke, backstroke, and American crawl. The instructors will evaluate the strokes down to the detail to ensure they not only know them but also use them in the most effective manner. As much as the candidates grow their physical and mental abilities, it takes much more to pass their special programs. It takes an attitude and heart, and heart can be more important than being a world-class athlete. At some point in their future program, each candidate will reach a breaking point. It's at that point when heart and attitude will determine whether they quit or continue at all costs.

* * *

In my experience as a counselor, I've learned that a lot of people lack self-awareness. People are unaware of their emotions and jump to being reactive, which exacerbates their situation. With fear, you must identify and accept that it exists and recognize what you're truly afraid of. Being overwhelmed with emotion and stress can be a signal that something more is bothering you than what's on the surface. Once you have awareness of fear, call it out, so that you have more control over it. Unawareness of the fear gives the fear control over you.

Fears range from avoiding an emotion to a full-blown phobia. In therapy, people can struggle with feeling sadness and anger, as an example, and often the root cause is from childhood experiences or messages that are carried into adulthood. It's important to break down the awareness: What am I feeling? Why am I feeling this way? What physical symptoms are manifesting from this feeling? When have I felt this way in the past?

* * *

After a few hours of early morning suffering, they rejoin their separate companies back in the barracks for the day's normal boot camp training. It's also worth noting that not all Dive Motivator candidates are going through boot camp. Others that have been in the fleet or recently passed "A" school can also volunteer to attend the training at RTC. "A" school or Accession training is where sailors receive technical training in their selected Navy Enlisted Classification. At the end of the two-month training the candidates must meet the minimum scores in a PST. Minimum scores are subject to change and differ slightly based on the special program requirements. The instructors overly stress that candidates should strive to achieve well beyond the minimum standards as the minimum won't get you through the actual course. Aviation rescue swimmer candidates must complete the following minimums in a single evaluation event: 500-yard swim in 12 minutes or less, 42 push-ups in 2 minutes, 50 curl-ups (sit-ups) in 2 minutes, 4 pull-ups, and a 1.5-mile run in 12 minutes or less. The RTC recruits that pass continue to their respective training. The recruits that fail to pass the PST get redesignated to a rating in the fleet's needs.

When I went through, AIRR candidates began their training with a four-week course at Naval Aircrew Candidate School (NACCS) in Pensacola, Florida. This is where Navy and Marine Corps officers and enlisted gain valuable education in flight physiology and survival of flight. AIRR now go directly to ARSS, and if they successfully complete the training, they then attend NACCS. This is to ensure the AIRR students maintain their high level of fitness from their prep training when going into the difficult Aviation Rescue Swimmer training. For the purpose of this book, I will keep the training order similar to my experience, but you can easily swap this and the next chapter for current accuracy.

During the first week of aircrew training, the candidates learn about the effects that gravity and high altitude have on the human body. This includes a practical phase where the candidates experience the altitude chamber to personally experience the effects of oxygen deficiency and learn methods of reducing panic and remaining calm in those conditions. This experience alone weeds out a few students in the pipeline, since not everyone can adapt to and function in such an environment. It's better to identify and acknowledge this early before putting themselves and others at risk.

Each morning begins with a muster and uniform inspection, then a fantastic breakfast at the local chow hall. In my six years served, this was the best dining experience I had, so I figured it was worth mentioning. After a hearty meal, the candidates spend hours on calisthenics, miles of beach runs, obstacle course training, pool and ocean swims, and water survival evolutions. (Ocean swims are no longer conducted at NACCS, as all water training is conducted in the new state-of-the-art pool facility.) The course spends a lot of time honing the student's ability to remain calm and survive in the open ocean. This consists of drown-proofing, learning efficient swimming techniques, working both as individuals and as a team, all while wearing full flight gear including boots. To graduate Aircrew training, each candidate must pass the following:

a. Be able to perform extensive daily calisthenics.

b. Pass a Navy Physical Fitness Assessment with a "good" in all categories for your age and gender.

c. Swim:

    1. 1 mile in flight suit in 80 minutes or less using sidestroke, breaststroke, or American crawl.

    2. 100 yards in full flight gear (flight suit, boots, helmet, gloves, and deflated life preserver) using each of the survival strokes

for 25 yards (sidestroke, breaststroke, elementary backstroke, and American crawl), followed immediately by a 5-minute drown-proofing (face-down prone float).

3. In full flight gear, tread water for 2 minutes followed immediately by 3 minutes of drown-proofing.
4. 200 yards (50 yards each using breaststroke, elementary backstroke, sidestroke, and American crawl).
5. Jump from a 12-foot tower and then swim 15 yards underwater wearing flight suit and boots using a modified breaststroke, immediately followed by floating using trouser inflation techniques.
6. 75-yard flight equipment swim using breaststroke only.
7. 100-yard swim using 25 yards sidestroke, 25 yards breaststroke, 25 yards elementary backstroke, and 25 yards American crawl.

The practical survival training involves classroom knowledge, which is then applied in the field. In the open ocean, the candidates learn safety techniques for signaling for help using different flares, radios, mirrors, and sea dye markers. They capsize inflatable life rafts in the ocean, then learn to work as a team, using their bodyweight as leverage to right them. They are also hoisted from the ocean by a Navy or Coast Guard helicopter to ensure they put their classroom and pool-simulated coursework to practical application. This is most of the candidates' first experience with the sounds and intensity created from the helicopter's forceful rotor wash. Rotor wash is the vertical air pressure caused by the revolving main rotors, which can well exceed 30 miles per hour of force. During different mission-sets, I've witnessed sailors being blown off ship decks, a sailboat capsizing in Australia, and the roof of a house in Korea being ripped completely off from the HH-60H's powerful rotor wash.

I was immersed in Aircrew training during the scorching summer of 1993. A Navy motor whaleboat staggered us individually 50 yards apart off the coast of Pensacola. We wore full flight gear, including flight suit, steel-toed boots, and an SV-2 survival vest with its corroded zippers, oversized helicopter helmet, and flight gloves. (The SV-2 has since been replaced with the updated CMU-33 and aircrew endurance vest.) It didn't take long to adjust to the warm temperatures of the Pensacola Bay as a Coast Guard HH-65 Dolphin helicopter approached for extraction. The high pitch of the main rotors and twin engines was ear piercing. The rotating blades pounded down a hurricane on the ocean's surface from the slow 70-foot forward hover. I recall the rotor wash feeling equivalent to having handfuls of gravel thrown in my face at close range. This unnatural sensation can and does cause panic in itself. We were instructed to push through the chaos and swim to the center of the storm directly underneath the hovering helicopter, which creates a slightly calmer pocket from the fuselage shielding the down pressure of the rotors. There we had to find the orange floating rescue strop connected to the rescue hook attached to the hoist cable being lowered by the crew chief. The large end of the rescue hook is then connected to our SV-2's harness-lifting V-ring. Once secured to the hook we swam back to pull the cable taut to help ensure we were clear of water hazards or debris. When safe, we gave the thumbs up as an indication that we were ready to be hoisted.

The crew chief depressed the up toggle on the hoist control, and I was dragged across the water until directly under the hovering Coast Guard helicopter. I was then raised straight up, and my saturated flight gear cleared the water. I distinctly recall smiling ear to ear, validating I was doing exactly what I was born to do. My nervous joy quickly turned to anxious fear as I looked directly down below and saw five hammerhead sharks circling where I was floating seconds prior. The crew chief stopped hoisting as I neared the cabin door and gave me a closed fist "hold" hand signal. The HH-65 helicopter aircraft commander then short-hauled me a few hundred feet away, where I was lowered back into the water. Ocean wildlife is one of those things you know exists but focusing on it won't make them any less dangerous. It'll just increase your heart rate and breathing, which leads to panic. And panic kills.

* * *

With fear there are mental and physical symptoms. Physical symptoms can include increased heartbeat, sweaty palms, headaches, gastrointestinal issues, shortness of breath, sweating, shaking, and in some circumstances disabling panic attacks. Mental symptoms can include thoughts of dread and doom, and irrational thinking. Emotions such as fear need to be unpacked to understand what's causing triggers. When you notice a physical response or surprising action, you start the process of awareness. At that point you can learn tools to face your fears. Better understanding allows you to do more to prevent increased anxiety and/or panic attacks. Learning breathing techniques and mindfulness helps ground you when you feel flooded with emotions.

* * *

A training week commonly referred to as "Disneyland week" because of all the mechanical devices includes a variety of terrifying survival techniques and practical education. However, it's a far stretch from the joys of Disneyland. Some of the training has evolved over time, which may slightly differ from my experiences in 1993. Plus, the training facility was demolished during Hurricane Ivan in 2004 and then rebuilt with more modern and advanced training equipment.

During my training, wearing full flight gear, the aircrew candidates were attached by their parachute risers and dragged face down back and forth across the pool. (After the new training facility was built, parachute drag exercises were eliminated from the curriculum.) When they reach the opposite side of the pool, the device dragging them flips them back over and heads the opposite way. Being uncontrollably dragged face down in the water is a daunting experience for anybody. The goal is to learn to remain calm and quickly flip themselves to their back and then find the quick release setting on each riser and disconnect the dragging chute. This is much easier said than done since the dragging device moves so quickly that by the time the student flips over and is searching for the quick release, they reach the end of the pool and are flipped back over. Some candidates will get one riser detached and then be dragged by the other riser sideways and completely submersed. Eventually they must learn to remain calm, push out the uncomfortable elements, and focus on the basics of righting themselves while simultaneously disconnecting the risers from the harness. A deployed parachute in the ocean can quickly become deadly, as it will either drag you across the surface of the ocean or fill with water and drag you down to your death. It pays to slow down, learn the basics, and develop the skills to operate efficiently in chaos.

Aircrew candidates also learn to escape from a parachute ballooned over them once they land in the water. Again, the dangers are very real and swift. A wet parachute will quickly suffocate you and drag you to the bottom of the ocean like an anchor. The basic techniques to escape from certain death aren't always complex, but the added elements and fear can prevent our ability to remain calm, focus, and avoid panic. Becoming paralyzed by fear when you have seconds to live is something the Navy worked hard to teach us to overcome. Those who couldn't learn to be comfortable when uncomfortable didn't pass, and in many cases they needed to be resuscitated on the side of the pool.

To aid in evacuating from a submerged helicopter crash in the ocean, the aircrew candidates train with mini SCUBA bottles called HEEDs. HEEDs were later replaced with helicopter aircrew breathing devices, which include a regulator like an actual SCUBA diving device. They fly with these small containers tethered to their survival vest, so they don't float away after water impact. The bottles offer a few minutes of compressed air to breathe, depending on various factors. The training consists of strapping the student in a five-point harness attached to a chair, connected to a rotating device located just above the pool waterline. The instructors then spin the chair completely upside down. Strapped into the five-point harness, disoriented, and with water rushing up their nose, the student must locate their HEEDs bottle in their SV-2 survival vest. They then position it the correct way since now up is down and down is up, insert the bottle's mouthpiece, clear it with a forceful blow, and take a few breaths. The aircrew candidate then calmly unstraps the harness and pulls themself from the submerged cage. Learning this technique in a controlled manner helps in the student's transition to the final phase of helicopter crash survival, the 9D5 helicopter dunker, better known as the helo dunker!

The helo dunker is exactly what it sounds like. Helicopters are top-heavy with their engines and main rotors centered atop the cabin.

As soon as it hits the ocean, it begins to sink as the main cabin floods with water. It then rolls to one side as the blades violently strike the ocean surface and rip off. The helicopter continues turning upside down and sinks to the bottom of the ocean at a rate of 9 to 13 feet per second. Since most Navy aircrew and pilots mainly operate over water, they train for both land and water mishaps. There's always a high risk of mishaps in the military when operating dangerous equipment in dangerous areas and situations, making this type of training imperative for increasing the odds of survival.

For those who are not helicopter aircrew, it's hard to imagine the scenario. Think about driving a van at night across a long bridge over a cold, deep lake when you hit black ice and crash through the guardrail. If you're lucky enough to survive the water impact, you try to process the urgency and magnitude of the situation as your body goes into shock from the freezing water. You take your last breath of air as the flooded car suddenly flips upside down and sinks like a rock. You have seconds to egress, and as you attempt to open your door, it is blocked and you realize the only possible exit is the right rear door window. In complete darkness you use your hands to find your way out as the pressure of the water depth is exploding your eardrums. That's a small glimpse into a comparable scenario minus the powerful rotor impact, other crew members, helicopter equipment, hazardous and possible burning fuel and debris, complex cabin egress paths, and ocean variables.

The 9D5 helicopter dunker, now replaced by the more modern modular egress training simulator, is basically a helicopter body minus the rotors, suspended by cables over a deep training pool. The cabin and cockpit are filled with aircrew candidates wearing full flight gear, which includes a flight suit, steel-toed boots, SV-2 survival vest, flight gloves, and a helicopter helmet. They are strapped into their respective seats with a locked five-point harness. The water survival instructor yells "Brace for impact!," and then they release the dunker. It drops from its suspended position in the air, hits the water, and then begins sinking. It floods with water as it sinks into the 15-foot pool. The aircrew, fastened to their seats, experience the claustrophobic and helpless feeling of the cold water rising from their boots, up to their legs, waist, chest, and neck. The training death trap then rolls upside down, with the crew suspended upside down in their chairs. Just like a helicopter flipping over from being top heavy with its rotors, you never know which way it will turn. Sometimes you roll forward, giving you more time to take a breath before going underwater. Sometimes you roll backward, which forces the water up your nostrils causing an instant suffocating response. The natural reaction is to unstrap and dive out the closest window, but that's a bad idea since there's a good chance the aircraft will roll over and either crush you or the rotors will chop you to pieces. I actually witnessed the aircrew prematurely exit during a crash while on deployment when a CH-46 Sea Knight helicopter had a sudden engine mishap during vertical replenishment. The aircrew dove out the back as the pilots autorotated into the ocean. Fortunately, everyone survived, but it would be an example of what not to do.

Before the instructors release the helo dunker, they order all the passengers to egress from a particular window or door. This doesn't give them much time to plan their strategy as they free fall to the water. There are primary exits for each crewmember based on their seating assignment, but they also need to be capable of finding an alternative if their primary egress point is obstructed. The instructors command the students to not panic, remain calm, and don't kick the person behind you. After the helo dunker sinks and takes a slow roll, the students must pause to ensure the turning ceases before considering their next move. It's an eerie moment of calm as the aircrew remains suspended upside down like drowning vampires.

The students feel for their five-point harness buckle located near their belly button. After a slight clockwise turn, the straps release, and then they right themselves. Hand-over-hand they feel their reference points to pull themselves through the cabin to the designated egress window or door. During their first of four helo dunker drops they will be instructed to egress their primary exit, which is relatively simple since gravity tends to suck you out the opening. It gets more complex when the aircrew are searching for a secondary exit. It's extremely disorienting being flipped upside down underwater, since up is down, down is up, left is right, and right is left. The students must remain calm, rely on their sense of touch, and have an abundance of patience and grace as others can and will unknowingly kick you in the face as they scramble to escape.

As you can imagine, not everyone remains calm in this training evolution, as it's very unnatural and panic inducing. That's the point! Rescue divers are always in the water in case of emergency, which happens quite often. The intense training is to prepare for an unthinkable emergency. And emergencies often happen within emergency training. Those students need to be pulled to safety, counseled, and rotated back into the next training class. Like most oxygen-deprived situations, the students must learn to remain calm and stay focused on their next steps. It's easier said than done, but the alternative is failing the course, or worse, sucking water, drowning, and dying.

Once the aircrew successfully egress a couple daytime crashes, they are graciously provided blackened goggles to simulate nighttime. It's difficult for most to remain calm when they have limited underwater vision, but it's next level once they put the blinders on. The students are completely reliant on touch as they feel for their safety harness buckle to release and then hand-over-hand pull themselves through the sinking coffin. Again, rescue divers are in the water to assist anyone in need, which in most cases counts as a failure of the training evolution. In a real-world scenario, they may have their HEEDs bottle to provide a few minutes of compressed air as they egress. However, for this training, they go without safety crutches to ensure they aren't reliant on compressed air to egress. Safety equipment can fail or get lost in the crash, making it crucial to be able to survive without it. Aircrew must confidently demonstrate the lessons taught to pass this crucial training.

Rescue swimmer candidates tend to find this training to be pretty awesome, but not all the aircrew students shared the excitement. Water is not a common element of comfort for most, but we felt at home. Otherwise, we were probably in the wrong profession. The helo dunker training alone could be the stopping point for future pilots or aircrew. However, if the students could remain calm and not panic, they could possibly egress and have the confidence to survive if an unfortunate mishap ever occurred in the fleet. I've spoken to several helicopter crash survivors and the common feedback is that this training was crucial to their survival.

In all cases, they said the water impact was much more violent than the training, but the egress mindset and techniques taught helped them survive the unimaginable.

On March 10, 1986, Charles "Chuck" Raygor was stationed at Helicopter Anti-Submarine Squadron, Light 34 (HSL-34) attached to the USS *Fahrion*, an Oliver Hazard Perry class guided missile frigate conducting operations in the Persian Gulf. HSL-34 flew the Kaman SH-2F Seasprite helicopter as the Navy's light airborne multi-purpose system. At the time, the Seasprite had enhanced anti-submarine and anti–surface ship threat capabilities. It was also versatile enough to be used for SAR, utility, and plane guard when attached to an aircraft carrier. On an overcast day in the Gulf, Raygor launched on a routine patrol in the Kaman with a crew consisting of two pilots, one passenger, and himself, an aviation rescue swimmer.

After operating miles northwest of the *Fahrion*, they returned to refuel before heading back out for a few more hours of flight qualifications. Frigates and smaller Navy warships (small compared to an aircraft carrier) have a helicopter landing pad and hangar on the aft of the ship. The rotary-winged aircraft approaches from the rear in a deck landing qualification. The aircrewman calls out the distance as they look out the opened cabin door, to ensure the wheels land safely in a painted circle on the flight deck. The pilots read their gauges, listen to the aircrewman, and look at the ground crew to help direct them in. Anything too far left, right, or back results in a crash in the ocean. Anything too far forward results in a rotor strike into the ship's superstructure. In less stable sea-states they will utilize RAST (recovery assist secure and traverse) landing. This is a tethering cable with a probe that is lowered to the flight deck and attached to another cable, which passes through a "beartrap" that houses a winch below the flight deck. The helicopter is then pulled in by the winch as the pilots maintain their level rate of descent. If an emergency occurs, they are able to jettison the RAST device cable and abort the approach. It can be a sketchy circumstance since you typically only use RAST in less than favorable conditions and everyone is highly stressed, willing to make a life-or-death decision within a split second.

The pilots lifted the fully fueled SH-2F and pivoted starboard off the *Fahrion*'s deck for forward flight. Not long after takeoff, Raygor recalled the eerie silence of both engines completely shutting down. Aircrew are accustomed to working in the constant extreme high-pitched noise of the engines and rotors. When that suddenly goes away, it doesn't take long to process what's happening. A catastrophic mechanical issue caused both engines to fail simultaneously. The pilots immediately informed the crew of the situation and got a "MAYDAY" radio call out to the *Fahrion* and other monitoring emergency resources. The entire crew automatically performed their overly rehearsed ditching procedures. Raygor says, "This was burned into every Seasprite aircrew. Every simulator flight ended with an immediate ditching situation and every flight at the Fleet Replacement Squadron (FRS) the instructors would randomly announce an immediate ditching situation sometime during it."

Passengers are briefed before each flight, but they aren't trained like aircrew, and even though ditching the aircraft is a possibility, the procedures aren't etched into their brain. Raygor made sure he and the passenger's harnesses were locked with seats up and aft. This allows the greatest response to the immense g-forces experienced in a crash. The HAC pilot dropped the collective control, causing an autorotation from 500 feet to the water. An autorotation is a helicopter emergency flight procedure used to force land during an engine or tail rotor failure. During the helicopter flare, where the nose of the aircraft is pitched up on the rapid descent, Raygor knocked out both cabin windows with a couple forceful strikes from his elbow and assumed the crash position. A flush of helpless adrenaline exited his body, providing a focused pause of realization. He remembers that unreal moment almost in slow motion as it unfolded, looking out his window at the calm, turquoise blue water quickly approaching. His eyes homed in on the slithering movement of all the striped and highly venomous sea snakes swimming on the surface. Then the SH-2F Seasprite landed, perfectly and peacefully on the Indian Ocean. A long second of silence washed through the crew as they contemplated their next procedures, which would require patience and discipline.

The floating helicopter bobbed on the ocean surface with no way for the pilots to control its destination. The rotors freely spun in neutral, naturally slowing their rotation as the pilots gently applied the rotor brake. The passenger, on his first flight ever, in shock and disbelief attempted to unstrap his safety harness and exit the still upright aircraft. Raygor authoritatively reached his hand over to prevent him from swiveling his harness's unlocking mechanism. The seasoned aircrewman informed him, "The ride isn't over!" The rotors continued to spin down and as they slowed, they began to tilt out of balance. The fuselage began bobbing and leaning more dramatically with each slow rotation. The crew was in a terrifying carnival ride, where they leaned closer and closer to their demise as the tips of the rotors got closer and closer to impacting the water. Then with one final rotation, the helicopter reached a 45-degree roll and the main rotors violently struck the water. The tremendous force of the rotor impact obliterated the blades off the main housing and knocked Raygor unconscious.

The Seasprite helicopter was taking on water as it sank and continued to roll. The violence abruptly ended, and the sounds of any remaining mechanical elements vanished. Silently they rolled to 90 degrees and descended into the darkness. As the seawater reached Raygor's neck, he suddenly regained consciousness. It took him a millisecond to process his predicament and that he was still alive and capable of survival. Taking a large breath of air, he then scanned the cabin and saw the passenger struggling to escape out the cabin door. The panicked passenger's helmet was snagged on something, and he fruitlessly thrashed the area with his hands. Raygor, still strapped in his chair sideways and sinking in the water, calmly unbuckled his own harness and reached up to help. He was able to free the snagged helmet internal communications system (ICS) cord and then forcefully pushed the passenger out the cabin door window to safety. He took one final deep breath before being dragged underwater.

Fully submerged in the dark of the helicopter cabin filled with cold, salty seawater, Raygor found handholds to guide his egress toward the cabin door. Halfway toward freedom, he felt a whiplash as he was pulled back. His survival gear was now caught on something, preventing his exit. With brute force, he pulled with all his strength but couldn't get free. Panic flooded his mind and his heart rate spiked. Morbid thoughts entered his mind about his parents and how they would find out their son died today. Knowing that panic kills, he immediately took back control of the situation. He shifted his thoughts from panic back to his helicopter egress training and calmed his emotions. This was before the HEEDs bottles were carried by aircrew, and he was desperate for air but needed to remain calm to elicit a decrease in arterial oxygen saturation to lower his heart rate. He focused on the situation and reasoned that one of his survival vest leg straps must have become entangled on his seat. Keeping one hand on the cabin door seal, he pulled himself back deeper inside the sinking helicopter. The pressure in his head felt like a vise as he explored in the pitch-black water, relying purely on touch. Hand-over-hand he followed his snagged leg strap down to where it was caught and peeled the nylon strap over the seat lock handle. Free of the obstacle, the helicopter silently sank away from Raygor as he floated through the open cabin door window. He reached down to the bottom of his vest and pulled the two inflation straps. The CO_2 cartridge fired, activating the inflation of the rubber bladders, and allowed the buoyant force to outweigh the water's gravitational force to assist in his ascent. As he moved toward the surface, he recalls the spookiness of the Kaman SH-2F Seasprite sinking to the black depths of the Persian Gulf. Raygor surfaced and took the biggest breath of air imaginable. The crew rejoiced in his escape and swam to meet up, where they connected their survival vest clips to one another. Raygor doesn't recall seeing any sea snakes near them, but honestly didn't think about them after the helicopter hit the water. The USS *Fahrion* launched a rigid hull inflatable boat (RHIB) and extracted the helicopter aircrew within fifteen minutes and delivered them back to medical.

Aircrew school focuses on both land and water survival. In land survival they learn to live off the land, build shelter, build a fire, navigate, and find food and water. I'll cover more in-depth land survival training during SERE School in chapter seven. Water survival made each candidate face their greatest fears from every facet imaginable in the water. The root of panic comes down to our inability to face our fears. This transfers into every aspect of our lives. On the flipside, those who claim to have no fear and are "fearless" have no safety gauge and put themselves and others at risk. Being afraid is OK; it's how we respond to that fear that defines us in the moment.

* * *

Journaling and talking to a trusted and safe person, plus being open to constructive feedback helps identify blind spots. It takes ongoing work to identify and respond to learned behaviors and triggers. Self-care, journaling, and mindfulness should become new habits for reducing fear. Once your nervous system is calm, this allows you to take those fearful thoughts captive and focus on things you have control over. Therapy is always an appropriate method to better understand ourselves, our triggers, and to learn tools for healing. The healthiest people tend to be those who seek therapy and continually learn to better themselves, which takes work.

* * *

Avoiding our fears is just as bad as having no fear; it's a total opposite end of the spectrum. The candidates who faced their fears, as uncomfortable as it was, were able to retrain their minds to navigate and thrive through the intense instruction. Charles Raygor and his helicopter crew were put to the test when they had to ditch their helicopter into the Persian Gulf. When faced with a moment of blackout and entanglement, Raygor was reminded of his training and calmed his fears to focus his efforts to safely egress the sinking aircraft. Everybody faces fear throughout their lifetime. Our ability to identify triggers and learn methods to control our responses leads us to healthier reactions to the chaos in our lives.

Chapter 3

Embrace the Suck

Hector Rodriguez, aviation rescue swimmer (retired), originally from Puerto Rico, spent several years stationed in Antarctica with Air Development Squadron 6 (VX-6). VX-6 was formed in 1955 at Naval Air Station Patuxent River, Maryland, among five other sister squadrons: VX-1, VX-2, VX-3, VX-4, and VX-5. Their mission was to develop and evaluate aircraft tactics and techniques. VX-6 flew both fixed-wing aircraft and rotary-winged helicopters in support of their mission-sets. Many of their flights were pioneering in unfamiliar and unforgiving areas of the world. In 1956, VX-6 became the first squadron with an aircraft to land at the South Pole with its ski-equipped R4D Dakota plane. Based in McMurdo Station, the squadron commonly called "Puckered Penguins" conducted operations in support of Operation Deep Freeze. VX-6's mission for Operation Deep Freeze was to conduct aviation operations and support for the U.S. Department of Defense with connection to the U.S. Antarctica Program. This included Task Force 43, flying missions to help construct airstrips, ice ports, and establish bases that enabled scientists to conduct geophysical studies on the frozen continent. Rodriguez was not accustomed to the cold temps growing up in the tropics, but he loved working in this foreign environment and putting his training to the test.

In 1979, Rodriguez earned the Air Medal for heroism during his tour in Antarctica supporting Operation Deep Freeze. He and his crew launched from McMurdo Station when they were alerted of a missing aircraft. His aircrew flew a UH-1N helicopter to search the area around

Mount Erebus on Ross Island, Antarctica. They were the first on scene, risking their lives in the frigid terrain to search for survivors of a New Zealand Air DC-10. Air New Zealand Flight 901 carried 237 passengers and twenty crew, on a scheduled Antarctic sightseeing tour. They planned to take off from Auckland Airport earlier that morning to spend a few hours over Antarctica, then return to Christchurch. Flight 901 took off, but it never returned. Due to a flight path coordinates change, the DC-10 was redirected from McMurdo Sound to a devastating path toward Mount Erebus, the sixth highest mountain on the continent at 12,448 feet. Sadly, all 257 souls perished in the crash. Rodriguez and his crew found and reported the horrific wreckage, altering the mission from rescue to recovery. Unfortunately, all missions don't result in positive outcomes. We train to save lives, but saving lives isn't always in our control. It's the harsh reality we must process and learn to live with, which no amount of training can truly help. The accident remains the deadliest accident in Air New Zealand's history of flight.

Years later, Rodriguez was stationed back near his warmer roots at Naval Air Station Roosevelt "Rosy" Roads, Puerto Rico. His squadron's main mission was supporting the Atlantic Fleet Weapons Training Facility. On a relatively average day of work, Rodriguez and his crew received a radio call to launch for a down aircraft. A Cessna Islander with ten souls had crashed into the waters somewhere between Puerto Rico and St. Thomas. The U.S. Navy SH-3G Sea King helicopter flew toward the last known coordinates of the crash site. Rodriguez sat in the helicopter starboard side door scanning the site for debris and assessing the situation for any possible survivors.

Four of the ten onboard went down with the sinking airplane. Six passengers remained huddled together on the ocean surface suffering from injuries and shock. The helicopter flew its SAR pattern into the wind, then slowed and descended to 10 feet and 10 knots. The crew chief tapped Rodriguez three times on the shoulder, and he jumped from the helicopter near the victims in the ocean. After giving an OK hand signal, he turned and swam toward the survivors. The HAC lifted the aircraft to 40 feet and backed away to the left to give Rodriguez room to perform his rescue. As the rescue swimmer approached the survivors, they all began frantically swimming toward him. At that moment, Rodriguez recalled his instructors saying that panicked survivors would come after you and try to "stand on your head." He immediately gained control of the situation and, with authority, communicated in Spanish for them to stop and remain calm, and he would come to them! They listened.

Rodriguez noticed four of the five victims were wearing inflated life vests, but a female passenger was not wearing any flotation. The dazed Cessna pilot had a large, open gash on his head, bleeding profusely into the shark-infested, tropical waters. Due to his injuries, Rodriguez grabbed the pilot first and towed him in a cross-chest carry toward the harsh rotor wash to the awaiting "horse collar" rescue strop. The pilot cried in pain as the hurricane force of the rotors pounding down on the ocean sent painful shards of saltwater at his injured face. The rescue swimmer worked efficiently to safely secure the survivor in the strop and keep him calm through the process. Rodriguez kicked backward with his fins to tighten the cable from getting fouled around their legs. He conducted a final safety check and gave a thumbs up to the crew chief. The hoist operator acknowledged the swimmer's signal, communicated to the pilots, and pressed up on the hoist control. The pilot dragged slightly across the water and then proceeded up safely into the helicopter, where the crew chief attended to his head wound.

As soon as the pilot was clear of the water, Rodriguez turned to head back toward the other crash victims. He prioritized the female passenger missing a life vest. She had a thousand-yard stare, shivering and suffering from shock. A thousand-yard stare is a blank, unfocused gaze seen on combat veterans or those experiencing extreme trauma. The phrase was coined after *Life* magazine published a painting in 1945 called *Marines Call It That 2000 Yard Stare* by Tom Lea. Rodriguez spoke in Spanish to calm her down, while he towed her toward the hovering helicopter. He strapped her in and signaled the crew chief for her to be hoisted up then returned for the other three.

One by one, he repeated the process to remove them from danger and get them safely to the SH-3G, hovering 40 feet above. Rodriguez recalls each of the final three victims panicking and trying to grab him, wanting to "stand on his head." Rodriguez used his strength and advanced techniques learned from Rescue Swimmer School to gain control of the situation. When he had control of each survivor, they calmed down to let him save their lives. For the final victim, he attached himself to the hoist with the survivor and locked his arms and legs around him as they were hoisted up together. He recalls the grateful victim crying, "Muchas gracias!" It's not often that you receive appreciation or even recognition for being a rescue swimmer. It's not the reason anyone volunteers for the job. However, on the rare occasion you receive gratitude it feels good, like you might have made a small difference that day.

When safely onboard the SH-3G, Rodriguez collapsed from exhaustion as the crew chief took over to perform medical checks and first aid on the six survivors. Rodriguez remembered the importance of remaining calm, never giving up, and learning to be comfortable in uncomfortable situations. The environment and conditions rescue swimmers work in are not ideal. Humans tend not to thrive in those elements, especially when attacked amid trying to save others. To save lives, we must embrace the suck.

* * *

Aviation Rescue Swimmer School is a sufferfest. There's really no other way to say it. Each day builds on the day prior, making every day harder than the one before. There's no easy day to look forward to other than the day after graduation, because the instructors will even smoke you on graduation. It is very intimidating, and you can do everything in your power to physically and mentally prepare for the training, but you'll never be truly ready for what's in store. The sooner you embrace the suck, the better your mind will be able to wrap itself around what's temporarily happening to your life.

In order to graduate from Rescue Swimmer School, the candidates must complete the course's technical and practical tasks, plus:

a. 90 minutes of intensive calisthenics and 30- to 35-minute cross-country runs daily.

b. 800-meter swim in 20 minutes wearing rescue swimmer equipment (mask, fins, snorkel, short wetsuit, and SAR harness with deflated flotation).

c. 400-meter buddy tow in 16 minutes wearing rescue swimmer equipment.

d. 2,000-meter swim in 50 minutes wearing rescue swimmer equipment.

e. 8 pull-ups in a flight suit and boots within 2 minutes.

f. Carry two 50-pound dumbbells 100 yards on flat terrain over four obstacles 12 to 14 inches in height within 2 minutes.

g. Walk 1 mile with a medevac litter within 16 minutes.

h. Swim 500 meters in SAR gear immediately followed by 400-meter buddy tow within 27 minutes.

i. Weekly strength training with free weights and machines.

b. Successfully complete CPR for professional rescuer.

The Navy Aviation Rescue Swimmer training location has changed through the years for various needs and reasons. The long-standing training facility I went through in 1993 was completely destroyed by Hurricane Ivan in 2004. It was replaced with a brand new 35,000-square-foot facility, containing a state-of-the-art pool and rescue equipment. The sight, sounds, and smell of the facility is intimidating in itself. It takes courage for each candidate to just show up each day. And that's when the suffering begins. For safety reasons, the physical training has progressed from what I will describe below of my experiences in the early 90s. The difficulty level has not changed, but the methodology has evolved to ensure safety is highly monitored. All physical training sessions are slated for three hours, which allows for two sessions plus forty minutes of running.

Each morning after uniform inspection the candidates quickly get dressed in shorts, T-shirt, and running shoes, then meet at the training sandbox or the concrete grinder outside the pool house. Equally spaced apart, they perform a stretch-set to avoid injury for what is about to come. Then they hit the sand, starting with push-ups forever. The instructor yells, "front-leaning rest position!" Immediately the candidates drop, locked in a plank position: legs, back, and buttocks perfectly straight, head raised and eyes looking forward. The rescue swimmer candidates remain in the locked-out position for twenty to thirty seconds. Everyone's arms profusely shaking, the instructor finally yells, "DOWN!" All go down and synchronously yell, "ONE!" And there they remain for another ten to twenty seconds with legs and back straight and nose touching the sand. "UP!" And they all go back to the raised, elbow-locked position. "TWO!" Twenty seconds later, "DOWN!" All go back down, "THREE!" You get the drill by now. Twenty seconds later, "UP!" And they all lock out, "ONE!" The count goes, 1-2-3-1, 1-2-3-2, 1-2-3-3, etc. Based on that logic each push-up is double the amount of the total count.

Whatever number they are targeting most likely won't come until several candidates collapse, which happens often and intentionally. Locking out a push-up for an extended period of time is difficult for anyone. We all have our limits. Each trainee's arms shake uncontrollably as candidates drop face first into the sand. As soon as one drops, loses count, or refuses to shout the number with authority, then the instructors start the count over. They have no hesitation in calling out the person or persons who messed up. They yell out their name, along with what they did, and shout, "START OVER, 0,0!" The rest of the group can either get angry at them or encourage them. Either way they'll be punished for any individual or group mess-ups. Over time, they will develop their leadership skills as well as how to work as a team. A common Navy phrase is a chain is only as strong as its weakest link!

With inoperably exhausted arms, the candidates roll over to get a nice coating of sand on their sweat-drenched bodies. They quickly find a partner and prepare for sit-ups. They alternate holding their feet for stability, which encourages less abdominal-focused exercise and more hip flexors and quads. The rescue swimmer instructors will order the group to "HOLD!" each sit-up at the halfway mark, working their abdominal muscles to something well beyond a chiseled six-pack. You can rest assured that the instructors will purposely trip the candidates up on the sit-ups and every count for that matter. The students may end up doing double or triple the amount they were expecting.

With abs blistering with pain, the candidates remain in the sit-up position as they transition to leg lifts and flutter kicks. Rescue swimmers need to perfect their flutter kick to give superior power and endurance in the ocean. A flutter kick has you lying on your back, hands flat pinned underneath your bottom, and straight legs hovering 6 inches above the ground with toes pointed out toward the horizon. With each count you raise one leg up, while keeping the other locked steady. With the next count you switch legs, which when in action makes a swimming flutter kick motion. Just like the rest of the exercises, flutter kicks use the same count format: 1-2-3-1, 1-2-3-2, etc. And to provide the candidates with the most opportunity for torture, the instructors pause with each kick until the student's legs shake uncontrollably. They then move to the next count and pause. Each flutter kick could take two minutes or more. At some point, someone will inevitably drop their legs to the ground, even if for a microsecond, and the rescue swimmer instructors joyfully start the count over. "START OVER, 0,0!" In the afternoon pool house training, the candidates will continue to do flutter kicks but while wearing full rescue gear, including heavy rocket fins on their feet. Other calisthenics like burpees are included in the morning routine to ensure every muscle group is thoroughly conditioned. The exercises are brutal, but the students are continuously reminded, "Pain is temporary, pride is forever!"

Many candidates drop the course in the first week purely based on the physical demands. They don't have what it takes physically, but more importantly they aren't able to evolve their thinking to overcome the discomfort of what the job entails. That's precisely why the instructors enforce such a difficult curriculum and environment to operate in. The last thing you want is for a rescue swimmer to quit a rescue in the middle of the ocean because they can't physically or mentally overcome the obstacles. One candidate in my class really struggled with keeping his legs up during the flutter kicks. The instructors were quick to single him out and even jokingly suggested that maybe his shoes were too heavy. They ripped his shoes off and threw them up in the trees then had us start over. Nope, that didn't help. He didn't make it much further in the program before dropping on request. He was a great guy with a positive attitude but didn't have what it takes to be an aviation rescue swimmer.

Most of the outdoor activities are conducted in sand, which ends up being one of the biggest unspoken obstacles. Sand dissipates a high amount of energy, due to its relentless resistance. Candidates burn more calories with an elevated heart rate, which accelerates the process of building strength and endurance. Sand also brings their level of grit for persevering in discomfort to an elevated and literal level. The candidates are continually covered in sand, called sugar cookies, which leaves irritating and festering grains all over their bodies. This friction is annoying and painful, but it forces them to push through pain while in constant discomfort.

By the time the candidates finish in the sandpit their arms and legs are Jell-O. They then move to the pull-up bars to continue building their upper body strength. Each strengthening exercise prepares the rescue swimmer for the durability required for performing rescue techniques in and out of the water. The students then transition to some cardio runs on the sandy beach and trails. The first week of training is brutal compared to anything most students have experienced. However, if they can persevere through the pain and mental game, then they will quickly transform into the physically capable creature the instructors are working toward.

Nutrition and hydration are an essential element of the fitness training. Each evolution contains water breaks, as the candidates are always at high risk of dehydration with their increased activities in the hot and humid Pensacola climate. At lunch, the team runs as a group to the base galley to refuel and then promptly returns to the pool house classroom to learn about rescue equipment, procedures, and advanced first aid. After a tough morning workout followed by a hearty meal, the candidates must maintain a sharp focus to retain an essential education, which they will be tested on. As much as they may want to nod off, they know better, as the instructors are watching and ready to pounce!

After the classroom it's time to get wet! The candidates wear full rescue gear to get used to operating in the essential rescue equipment. For me, this included short wetsuit top, harness, SAR-1 flotation, rocket fins, mask, and snorkel. Today, the only differences are helmets and the new and improved integrated harness and flotation TRI-SAR, which provides better support for hoisting among other benefits.

Starting with pool conditioning, the students quickly learn that everything is a competition. They swim the 25-yard length of the pool underwater and then sprint back using the American crawl on the surface, commonly called "under overs." This exercise is repeated over and over. The students must dive deep, remain calm, and control their breathing. They are taught that most things come down to staying calm and controlling their breathing. The instructors warn the candidates that if they try to swim underwater too close to the surface, they are at greater risk of experiencing shallow water blackout. This is a result of hypoxia or low oxygen levels to the brain, which causes the swimmer to faint in the water. When this happens during training, all exercises immediately stop for the emergency so a safety instructor can attend to the victim. The last few swimmers to complete the exercise are graciously rewarded with doing it again while the rest of the class watches and rests from the side of the pool. This was a difficult exercise for some, which led to their DoR during the first few days of training. (Under overs have since been replaced with four total underwater sprints allowed during Swimming Proficiency and Conditioning, and the swimmers can come up for a breath at any point during two of the four.)

The candidates must learn to be meticulous about their attention to detail in every task. I'll discuss this more in chapter six. Their masks and fins need to be stored and carried a certain way, which over time becomes habit. They are the rescue swimmer's lifeblood in the water. Although a rescue can be executed without them, they are key elements to providing a competitive advantage over an aggressive victim, strong ocean currents, and the harsh power of the helicopter's rotor wash.

During my training in 1993, my entire class lined up for pool training following our after-lunch classroom coursework. We immediately noticed all our gear was missing. One of the students didn't store their gear correctly, or, more likely, it was just another mind game. Either way, all the gear was in a large pile at the bottom of the 15-foot-deep pool. The instructors yelled, "What are you waiting for!" Without hesitation we all dove in and headed deep toward the blurry black pile. We tried sorting through to find our personal gear, which had our last name stenciled on it. It was total chaos with over twenty rescue swimmer candidates deep in the water, trying to read each pair of gear to find their own two fins, mask and snorkel, rescue harness, and SAR-1 flotation vest. If finding the gear wasn't tough enough, we also had to don the gear underwater, which added another layer of complexity. Anyone surfacing for air was welcomed with shouts to "DIVE!" from the instructors. This test filtered out some candidates who failed to remain calm underwater while focusing on the task as they were repeatedly kicked in the face by fellow desperate trainees.

Compartmentalizing fear and focusing on remaining calm became very real during one particular training evolution. The aircrew water instructors rolled the individual candidates up in a parachute and then wrapped the shroud lines tightly around them. We were then pushed into the deep water. If this doesn't induce panic, then I'm not sure what does. We had to face our fear of drowning, remain calm, control our breathing, and be patient. The technique to escape is quite simple if you have the discipline. With your arms pinned to your side in the tightly wound parachute cocoon, you gently wave your hands back and forth. This creates a small air pocket in the water-saturated parachute. As your body remains still, you slowly start gliding out. If you panic and thrash around, then you suck water, black out, and most likely die. Of course we had safety instructors there to ensure we didn't die. Human nature is to panic, which is why we learned in extreme scenarios to trust the techniques as well as our own ability to overcome the impossible.

A rescue swimmer's main method of transporting victims through the water is to tow them in a variety of different techniques. The candidates conduct these methods often to perfect the technique and transform it as a natural physical skill. They swim 200-yard buddy tows, where they partner up and use a cross-chest carry to tow the student back and forth across the pool. They then alternate as the victim and get towed, which is a nice momentary rest. Until they switch back and do it all over again.

The candidates conduct mile swims both in the pool and in the ocean. During one of my mile swims in the Pensacola Bay, there was an unusually high infestation of jellyfish. Every stroke I took, a jellyfish hit into my mask, stinging me all around the rim. With each reaching stroke, my hands cupped jellyfish and I pulled them back. My fellow students and I were swollen, red messes. It was a day we all probably wondered what we were doing there but fully embodied "embracing the suck," as that brief day of misery would pass.

* * *

The description of SAR training gives an example of how in a period of time one might need grit to get through the experience. However, in counseling, often people come in with difficult life experiences and have either built grit into those experiences or have not. Wikipedia defines grit *as: a positive, non-cognitive trait based on an individual's perseverance of effort combined with the passion for a particular long-term goal or end state.*

* * *

On April 7, 2003, the aircraft carrier USS *Harry S. Truman* was conducting side-by-side resupplying with the USNS *Spica* in the Mediterranean Sea. They, along with another carrier battlegroup, were steaming toward the Persian Gulf in support of the U.S. invasion of Iraq. The *Spica*, a merchant supply ship, essentially came alongside the *Truman*, matching its course and speed. Ship personnel then shot lines across to the other deck, which are then secured to allow supplies to be hauled from the merchant ship to the carrier across the anchored lines. Additionally, helicopters attached to the carrier conducted vertical replenishment from the supply ship to the carrier. In this case, a helicopter attached to the supply ship also assisted in VERTREP operations. They transferred heavy cargo wrapped in nets suspended by cables attached to a cargo hook centered beneath the aircraft. During that moonless night, the carrier flight operations completed earlier and Lieutenant Pastorin, Lieutenant Prikarsky, AW2 Tracey Hoff, and AW1 Jeremy Burkart from the "Dusty Dogs" Helicopter Anti-Submarine Squadron 7 (HS-7) were assigned to the twenty-four-hour Alert 30. Each day, a new crew is assigned to Alert 30 status, which means the crew prebriefs and remains ready to launch within thirty minutes when a distress call comes in.

The USNS *Spica*'s Aerospatiale SA 330J Puma helicopter, piloted by Richard Good and Richard Budd, was flying between ships when it fell from radar and radio contact. The HS-7 Alert 30 aircrew received the emergency call and they launched within nineteen minutes. A nearby aircraft carrier, USS *Theodore Roosevelt* and Arleigh Burke class destroyer USS *Winston Churchill* also launched rescue helicopters. Good and Budd were in the frigid 55-degree water with a strobe light blinking and sea dye marker, which created a green glow around them. HS-3 plane guard helicopters from the *Roosevelt* arrived on scene first. They immediately noticed all the fuel and debris in the water. Lieutenant McKechnie and Lieutenant Smiley brought the HH-60H helicopter into a hover near the survivors and AW1 Greg Baker lowered Rescue Swimmer AW3 Jason Boutwell down 70 feet to the water.

The civilian pilots weren't wearing the standard equipment that military pilots typically wear, so Boutwell needed to treat them as free-floating survivors. Military pilots are required to wear a survival vest with a harness, which includes a lifting device for rescues. Since they were wearing civilian clothes without a lifting device to expedite hoisting, they would require the rescue strop secured around their back and under their arms. Boutwell communicated with the survivors and then prioritized rescuing Richard Good first, who was without a helmet or strobe light, which would make him more difficult to find if he drifted away. Boutwell went through the proper checks, freeing Good of any potential hazards and then signaled for extraction.

While Good and Boutwell were being hoisted to the HS-3 helicopter, the HS-7 Alert 30 HH-60H was preparing to lower AW1 Burkart to rescue Richard Budd, who had drifted 300 yards from Good. Sitting in the cabin door, wearing his SAR gear consisting of a full wetsuit, mask, snorkel, fins, harness, and flotation, Burkart broke a chemical light and shoved it in the small holder atop his mask. The chemical light provides minimal visibility for a night rescue without disrupting the swimmer's night vision but, more importantly, allows the crew chief to maintain visual contact with the swimmer in the water. As Hoff hoisted him down 70 feet to the black ocean below, Burkart instantly felt a gag reflex from the overwhelming smell of fuel in the water being kicked up by the strong rotor wash.

"As soon as I got into the water," Aviation Warfare Systems Operator 1st Class Jeremy Burkart of HS-7 reflected, "I was lowered into an oil slick. I was covered in jet fuel from the helo and I couldn't see anything through my mask."

He placed his mask and snorkel on his forehead since it wasn't helping but actually impeding his search for the survivor. With 3-to-4-foot swells, he lost sight of Budd and tried calling for a vector to the pilots via his legacy AN/PRC-125 radio but couldn't hear anything, even if there was a response. Burkart decided to kick hard to lift his body higher over the swells, which worked as he spotted the downed pilot's strobe light reflecting on the dark ocean surface. The rescue swimmer swam toward Budd, mostly with his eyes closed due to the pain caused by the fuel and fumes. When he reached the pilot, he pulled out a rescue swimmer classic greeting: "I'm a rescue swimmer!" he shouted above the noise, "I'm here to help!" Burkart then meticulously went through his well-practiced disentanglement checks. Budd was pretty banged up but had no suspected spinal injury, which would require a rescue litter. He was OK to hoist up, so Burkart signaled to his crew chief, Hoff.

Both helicopter squadrons flew the Puma pilots to the USS *Harry S. Truman*, where they were brought to the carrier's hospital to recover. The HS-7 aircraft's survivor and swimmer were so dowsed in fuel and oil that the cabin of the helicopter was an oil slick, and the pilots thought they had a fuel leak due to the pungent smell. Burkart's rescue swimmer equipment was deemed unusable after the save, and he went the remaining months of the deployment as a dry hoist operator as his custom wetsuit and gear were being replaced back in the United States. Because of his long exposure to the JP5 fuel in the water, his eyes felt like fire and were almost completely swollen shut. Although they would eventually heal, they never truly were the same. The damage experienced that night saving a life seemed to accelerate his vision decline, which he still deals with today. Burkart was awarded the Navy and Marine Corps Commendation Medal for heroism. Due to the timing of the rescue, days prior to the invasion of Iraq, the carrier had scores of media and press onboard. While trying to recover from his injuries, Burkart was called to take several interviews about the rescue, which is extremely uncommon since most Navy rescues fall under the radar due to the sensitivity of the missions and secrecy of those involved.

* * *

Those who have built grit into their normal lives are able to take difficult setbacks in life and use that to continue to propel them forward. In hard times, it's perseverance that gets you through. If you see quitting as an option, then, more often than not, that's what you'll choose.

* * *

In August 2005, a massive Category 5 Atlantic hurricane named Hurricane Katrina destroyed much of New Orleans and the surrounding areas. It originated as a tropical depression, which intensified as a tropical storm as it moved west toward Florida. Two hours prior to landfall, Katrina strengthened into a hurricane, wreaking havoc on the south end of the state. The hurricane weakened back to a tropical storm as it moved farther west, but then rapidly increased its intensity back to a Category 5 as it merged with the warm waters of the Gulf of Mexico. By the time it impacted southeast Louisiana and Mississippi, Hurricane Katrina had weakened to Category 3. Winds of 120 mph and 15 inches of rain caused consequential damage to the area, but the most significant destruction came from flooding. Due to engineering flaws with New Orleans levees or flood protection systems, 80 percent of the city was underwater. The levees experienced fifty-five breaches, according to the Army Corps of Engineers (COE). The COE began immediate operations to repair the damaged levees, pump out water, distribute food and fresh water, and remove trash and debris. The catastrophic storm caused 1,392 fatalities and over $125 billion in damage.

Eighty percent of the people had evacuated, leaving fifty- to sixty thousand people in their homes or seeking refuge in the Superdome NFL football stadium. The U.S. military immediately responded by coordinating assistance from all branches. They sent eighteen thousand active-duty service members alongside forty-three thousand National

Guardsmen for relief efforts. The U.S. Coast Guard was first on scene, followed by the National Guard. Coast Guard helicopters were scattered all over the city rescuing victims from rooftops as the water continued to rise. Panic was spreading through the city as people lost their homes, relatives, friends, pets, and hope. They didn't have clean water, food, communication, or plumbing. Temperatures consistently remained above 100 degrees Fahrenheit with matching humidity. The lack of fresh water, increased anxiety, and alcohol consumption led to major dehydration in the victims. Violence, looting, and other acts of desperation took place. The military rescue crews weren't just dealing with trying to save lives, but also having to work crowd control as hostile victims fought to be evacuated.

Learning victim and crowd control, plus being authoritative in all situations, was a key element of Rescue Swimmer School. Rescue crews will always try to prioritize based on injury, age, and severities, but in the aftermath of Katrina, several angry individuals tried to force their way past critical victims, elders, women, or children. The aircrew being lowered down in the center of the mob had to enforce their authority and risk injury to filter through desperate people to hoist the priority victims first.

The U.S. Northern Command established a Joint Task Force Katrina (JTF) in Camp Shelby, Mississippi. JTF Katrina was the military's primary on scene command for the disaster recovery efforts. Air Force pararescue (PJ), Navy aviation rescue swimmers, Coast Guard helicopter rescue swimmers, and aircrew from the Army, Army Reserve, Army National Guard, and Marine Corps coordinated flight efforts for rescues and logistics. All services flew around-the-clock helicopter rescue missions, loading their cabins to full capacity to deliver survivors to dedicated triage areas outside the flood zone. During the height of rescue operations, there were more than 350 helicopters and over seventy fixed-wing aircraft engaged in the Katrina relief efforts. With all of the swarming aircraft, the scene was compared to a highly active hornets' nest—complete chaos of professionals risking their own lives to save others. Fortunately, there were minimal aircraft mishaps as the crews were diligent in their flight safety protocols. The U.S. Coast Guard air, boat, and ground units would end up rescuing over 33,500 personnel. They were later awarded the Presidential Unit Citation.

The Navy's USS *Bataan* was already operating in the area and was able to launch two MH-60S from Helicopter Sea Combat Squadron 28 (HSC-28) and four MH-53 from Helicopter Mine Countermeasure Squadron 15 (HM-1) for immediate SAR operations. Additional ship and helicopter assets were sent from Norfolk, Virginia, including USS *Iwo Jima*, USS *Shreveport*, USS *Tortuga*, USS *Harry S. Truman*, and USS *Whidbey Island*. The increased demand for helicopters led to the deployment of aircraft from across the country. For instance, Helicopter Anti-Submarine Squadron Light 43 (HSL-43), HSL-47, HSL-49, and HSC-21 arrived with multiple MH-60S Seahawks from NAS North Island, California. Three Marine squadrons from MCAS New River, North Carolina, sent six CH-53E Super Stallions and two CH-46E Sea Knights, and reserve Heavy Helicopter Squadron 772 (HMH-772), from Willow Grove, Pennsylvania, sent four CH-53Es. Naval Aviation units also provided crucial logistical support including Fleet Logistics Support Squadron VR-57 and VR-58. The Logistics Squadron moved in construction battalions or Seabees and HSL aircrew to focus on evacuations and transporting supplies.

Aviation rescue swimmers Tim Hawkins and Scott Chun, assigned to HSL-40 (redesignated to HSM-40), launched to patrol over the flooded streets in search for those in need. For safety and sanitary reasons, the Navy aircrew wore latex gloves under their standard flight gloves. That coupled with full flight suits and survival gear made for miserable working conditions. Hawkins recalls being drenched in sweat from head to toe and how difficult it was to work with wet, slippery latex gloves. Aviation rescue swimmers typically operate over the ocean for maritime rescues, but they are trained to adapt to whatever need is required and push through uncomfortable elements to save lives. This was certainly the latter circumstance. They didn't have time to overthink the scenario, since time was crucial to evacuate people from the flooding waters. With no power in the city and neighborhoods and the water rising to the rooftops, it seemed like something from an end-of-times fictional movie. As the SH-60B Seahawk helicopter slowly crept over the eerie residential areas, the rotor wash created a wavelike current through the flooded streets. It was a surreal scene searching through the apocalyptic landscape.

During a flight over a suburb of New Orleans, the aircrew spotted a person in an open window waving a white blanket in distress. The pilots brought the helicopter into a 70-foot hover over the three-story building. The water was rising past the first floor, forcing the occupants to find higher shelter. Acting quickly, Chun lowered the rescue hook to Hawkins who connected his harness-lifting V-ring to the large hook. He released the gunner's belt secured around his chest and Chun, using the hover trim control, raised the rescue swimmer up out of the helicopter and then down to the rooftop. Hawkins disconnected from the hook, gave a thumbs up to his crew chief, and went to work investigating the structure of the building. He needed to find his way down to the second floor, where the victim was waving the blanket for help. He scanned the area and fortunately found an aluminum ladder collapsed on the floor, which he extended and secured to climb down a couple floors.

Hawkins cautiously moved inside the building and searched for the people requiring help. He recalls experiencing a gag reflex due to the intense smell of hot sewage and rotting death. He found a frail elderly lady resting in a wheelchair near the window, with a white blanket laid across her knees. She was emotionally and spiritually overwhelmed to see the aircrewman enter her home. She was raising her hands up to the sky, exclaiming, "Thank you, Jesus! Thank you, Jesus, for sending an angel!"

Hawkins gently calmed the grateful woman down and communicated his intentions for getting her to safety. That's when he noticed an elderly man lying lifeless on a nearby bed. Human feces were piled and smeared everywhere, causing Hawkins to vomit in his mouth a bit. At first glance, he assumed the man was dead. He carefully navigated through an obstacle of clothing and trash to properly assess the victim. Hawkins initiated physical contact by grasping the man's shoulder and talking to him to gain a response. To the rescuer's surprise, the man jolted straight up in a daze. Hawkins had to calm himself after being startled as well as the newly risen man. He explained why he was there and then conducted a medical assessment to check the man's vitals and identify any injuries that could prohibit his evacuation.

Hawkins carefully lifted the frail man and carried him to the base of the ladder leading to the rooftop. He had to reposition the man in a way that would allow Hawkins to carry him up the long ladder. The man's weight sagged from gravity as Hawkins placed him in a fireman's carry over his right shoulder. Rung by rung, the rescue swimmer slowly edged their way to the roof. Hawkins's clothes were saturated with sweat as he pushed through the miserable elements. The helicopter engines and rotor wash were ear piercing, making it difficult to communicate with the victim. The high wind generated from the rotors provided some much-needed cool air as Hawkins securely held the victim.

Protecting the man's face from the scattering projectiles created by the hovering helicopter, Hawkins signaled Chun to lower the rescue hoist. The SH-60B crew chief, hovering 70 feet above, lowered the hoist and rescue strop. Hawkins made sure to stay clear of the electrostatic discharge created by the high-speed rotors, allowing it to ground out on the rooftop. He then quickly went to work, attaching the rescue strop securely around the victim. The rescue strop is placed under the arms and across the back and then connects into the large rescue hook. There's also a safety chest strap to prevent the person from sliding through as well as a safety strap to pin the arms down. The arm safety strap was damaged from a previous rescue and depending on the scenario the risk of the victim sliding through could be low. However, due to the elderly man's frail condition and the fact that they were hoisting over buildings rather than water, Hawkins decided to ride up with him. He connected his SAR harness-lifting V-ring to the hook and wrapped his arms and legs securely around the man as they hoisted up. Once they reached the helicopter cabin door, Chun safely pulled them onboard, disconnected them from the hook, and secured the man inside the cabin. Hawkins then reconnected to the rescue hook and was lowered back down to retrieve the helpless woman in the wheelchair.

The rescue swimmer reached the rooftop and detached from the hook. Chun hoisted it back up to avoid getting fouled on a surrounding hazard. Hawkins climbed back down the ladder and reentered the building to find the patiently awaiting woman. Her big brown eyes and smile were infectious. With no words, her eyes told a detailed story of her deep excitement and gratitude to be rescued. The rescuer let her know that it may be uncomfortable, but he was going to need to carry her to the roof to be evacuated. Hawkins carefully placed her in a fireman's carry, draped over his right shoulder, and carried her through the maze of hoarded stacked items and out to the ladder. She was much lighter than the man, which made the climb slightly easier and faster as he made their way to the rooftop. As he stepped off the top rung, he gave a thumbs up to the awaiting crew chief.

Chun lowered the hoist and Hawkins repeated the previous process of securing the victim. However, since she couldn't stand on her own, he had to overcome some weight and balance challenges to ensure she was strapped into the rescue strop. Everything looking good, he gave Chun the thumbs up hoist signal. The crew chief pressed up on the hoist controller and she began lifting toward the helicopter. In that split second, a warm flush of panicked adrenaline overcame Hawkins when he realized he hadn't connected himself to the rescue hook. He visualized her in slow motion, escaping from his safe controls. With the arm strap damaged, there was a potential the disabled woman could slip through the strop to her death. Not on his watch!

This concern was top of mind for all aircrew, as days prior a victim had slipped out of the strop of another helicopter and fell 40 feet to her death. Without hesitation or regard for his own safety, Hawkins jumped up and grabbed hold of the top webbing section of the rescue strop. Hawkins held on tightly with his gloved hands in a "locked-V" grip, while securely wrapping his legs around the disabled woman as they were hoisted up 70 feet at a rate of 215 feet per minute. The hoist motor automatically decelerates to 50 feet per minute when it reaches 10 feet from full up to the hoist housing. This dramatic slowdown causes a whiplashing jerk, which caused Hawkins to slide farther down on the strop. He was hot, sweaty, and amped with adrenaline, which is a hormone that makes your heart and lungs work faster by sending additional oxygen to your muscles. This can result in a temporary boost in strength, which is exactly what Hawkins needed in that moment. Chun, meticulously monitoring the two moving up the wire, observed the rescue swimmer was hanging about 2 feet lower than he normally should be. He didn't notice any panic in Hawkins's face or any cause for concern, so he continued hoisting them up.

Hawkins's fingers and forearms were burning with the fire of fatigue as he locked in his kung fu grip. At this point he was fully committed, as it was life over limb. Any slip would result in an instant game over for him as the helicopter was hovering high above houses. When they finally reached the SH-60B's cabin, Hawkins yelled to alert Chun that he wasn't attached to the hoist. Chun quickly processed this new information but remained calm to safely get them both onboard. He paused, hoisting long enough to allow Hawkins to get a foothold on the helicopter's starboard side tire. Chun then instructed a photographer who was riding as a passenger to grab Hawkins's hands and pull him in as he simultaneously lowered the hoist and pulled the woman in. In one fluid motion, they were both safely in the cabin. Chun quickly slid the cabin door closed and secured the helicopter cabin for forward flight. The pilots acknowledged his status update and set course to a local triage center. Hawkins laid on his a back for a few seconds to catch his breath, and then he and Chun commenced administering first aid to the survivors.

The U.S. military flight crews operated around the clock, only breaking for forced required downtime, maintenance, refueling, and meals. Hurricane Katrina brought more rescues in a few days for all involved than most would normally get in a full career. The rescue swimmers and aircrew train hard for moments like these, but it's in these moments that they are pushed beyond their perceived limits. Most overwater rescues take twenty to thirty minutes, where the Katrina rescues were continual and took the better part of a day. During the chaos of trauma, things don't always happen as planned and mistakes can occur. Hawkins was faced with a complex rescue of a disabled, wheelchair-ridden woman being hoisted in a rescue strop rather than a contained basket, which unfortunately they didn't have. When he realized she was about to head up alone without the additional protection of the arm safety straps, he committed life over limb to ensure her well-being. That's the type of attitude and response that separates the military's aviation rescue swimmers from the rest.

* * *

In this example, the rescue swimmer was dedicated to doing the job he trained for, no matter the cost. Life is hard. How are you going to approach those setbacks? Are you going to persevere and figure out a way to get through it and learn from the opportunity of a setback? Or are you going to see things as a failing and be stuck in a fixed mindset? Grit along with mindset is something that we can thankfully learn. Obviously through the examples given here, there are many areas where grit can apply to your own life; it just requires your determination.

* * *

Aviation Rescue Swimmer School requires intense physical and mental aptitude. Some of the toughest built athletes don't make it past the first week, while many unassuming candidates find their inner strength to push beyond their physical limitations and graduate with honors. The instructors teach the candidates lessons about working as a team as well as pushing their bodies further than what they had ever experienced. The student's perceived limits are stretched each day as their mind and body significantly rewrite their baseline of potential. We all knew it sucked to be out there getting physically and mentally hammered. As soon as we accepted this as our new way of life, which by the way we volunteered for, we learned to focus on getting through each difficult evolution, one at a time, while building each other up. Each step forward was a step closer to graduating, and, eventually, we would survive the sufferfest.

Chapter 4

Fight, Flight, or Freeze

In 1986, the Essex-class aircraft carrier USS *Lexington* was operating at night in the Gulf of Mexico when it received a distress call from a U.S. Coast Guard dispatcher. Aviation rescue swimmers Bill Gibson and Rick Williamson, assigned to Helicopter Combat Support Squadron 16 (HC-16), launched in an SH-3D Sea King to investigate. The pilots hustled the rescue helicopter toward the coordinates provided as Gibson and Williamson donned their wetsuits and rescue gear and Crew Chief Ed Howard rigged the cabin for a night rescue.

The initial report suggested anywhere from twelve to twenty people in the water from a capsized boat. The helicopter aircraft commander flew a box search pattern using their spotlights while all aircrew scanned the dark horizon looking for signs of life. The left-seat pilot then miraculously spotted a strobe light flashing a mile away at the eleven o'clock position. The HAC acknowledged and immediately pulled the heavy aircraft to the port. With limited intel and not knowing what they would find, they prepared in their minds for all possible scenarios including mass casualties.

The SH-3D passed at an altitude of 150 feet over an orange life raft with four souls onboard as the crew chief tossed out a MK-58 MOD 1 smoke to mark their position. The pilots then continued a wind-line SAR pattern to circle back to come into a hover to deploy the rescue swimmers. The co-pilot called the rollout as the HAC brought the Sea King to wings level and engaged the approach as the two rescue swimmers anxiously sat in the cabin door. Williamson grabbed two chemical lights, or chem lights, from the helicopter's rescue bag. He snapped them both, causing the chemical reaction within the plastic that produces a green glowing light. They both slid the chem lights into a small round holder attached to the top of their masks to help provide light when in the dark water and more importantly for the crew chief to not lose sight of the swimmer in the pitch-black night. At 100 yards from the raft, they established a creep at 40 feet and 10 knots until on top of the survivors. The HAC engaged the hover and the crew chief lowered Gibson and Williamson 40 feet down into the dark abyss. Once the swimmers were clear from the rescue hook, the HAC moved the hovering aircraft back and left to reduce the amount of rotor wash for the rescuers to do their work, as it had already pushed the life raft a hundred yards away.

The rescue swimmers swam up and down the high ocean waves, losing sight of the raft on each down swell. The crew chief watched the glow of the swimmers' chem lights as they topped each crest, searching for the drifting raft. Howard informed the pilots of the difficult conditions for the SAR team in the water, and they meticulously trained the helicopter's spotlight toward the bobbing life raft. Gibson and Williamson simultaneously kicked toward the light's direction to eventually reach the stranded victims where they established verbal contact.

Gibson asked the four survivors if there was another life raft or anyone else missing from the capsized vessel, since the initial report suggested several more victims. Fortunately, it was just the four of them. They communicated how the rescue would be conducted to set expectations and timeframes with the victims suffering from shock. Gibson then pulled the oldest person into the water and gave him a brief moment to adjust to the water and calm down. He then locked him in a cross-chest carry and towed him toward the helicopter to be hoisted up via the rescue strop by Howard. Gibson was quick to place the horse collar around the victim, attach the safety straps, and clear him of any possible obstructions. The rescue operation was pretty textbook as Howard helped reduce the gap by guiding the pilots to move the helicopter closer to those ready to be hoisted. The rescue swimmers alternated as Williamson, towing another survivor to be hoisted, passed Gibson who was returning to the raft.

Gibson encouraged the third survivor to enter the water and informed the final one in the raft that Williamson would be back shortly to bring him in. He then towed his survivor toward the rotor wash of the hovering SH-3D, passing by his partner heading back to the life raft. Gibson entered into the circle of thrashing water and gave a thumbs up, and Howard acknowledged by lowering the rescue strop down to his two o'clock position. Gibson recalls looking up through the hurricane-force winds of the helicopter's main rotors and noticing the dangling strop stopping about 30 feet above him. Next thing it was being hoisted back up as the helicopter shifted to the left and away from him and his survivor. For a brief moment he had no idea what was going on as he processed the possibilities. Was there an engine failure? Were they low on fuel? Was there a hoist malfunction? Then a horrified Gibson turned to his left.

The rescue swimmer, holding a boat wreck victim and treading water in the pitch dark, looked left and saw the unthinkable. A 400-foot cargo vessel about to crush them both! Without hesitation, Gibson grabbed hold of his survivor and intensely began flutter kicking to the right of the fast-approaching ship. They were both being sucked under the massive keel as he swam with all his strength. As the cargo ship steamed forward, it created a massive wake of waves. Gibson pulled his survivor up toward the crest and then would be sucked back down toward the trough. He found the best rhythm of survival possible as he fought up and down the wake of the entire length of the ship.

Gibson recalls how insanely arduous it was to continually fight against being sucked under and crushed. It would be difficult to survive this alone, but carrying a survivor was next-level impossible! The exhausted rescue swimmer saw the end coming near as they reached the aft of the unrelenting vessel. He heard and felt the haunting, massive three-story-high propellers violently chopping through the night ocean as they were uncontrollably pulled toward them. He yelled to his survivor, "Take a deep breath! Kick as hard as you can and don't stop!"

The rescue swimmer took a final breath of air and pulled the survivor underwater. He frantically flutter kicked deep down into the pitch-black ocean as the water-depth pressure crushed their ears. The deafening pounding of the propellers reverberated through the water. The deeper he went, he kept thinking, "I'm not hit yet. I'm not hit yet. I'm not hit yet." In the midst of chaos, his life was eerily calm as he knew it wasn't if, but when they would be hacked to pieces. He just kept kicking. He would never stop kicking. Then everything went completely silent and calm. They had both astonishingly made it down and under the cargo ship's propellers. Moments later they surfaced on the other side of death in the ship's wake. They were alive!

As the ship continued its coordinates with no knowledge of what occurred, Gibson immediately vomited the gallon of seawater he consumed during his underwater tow. He and his survivor treaded water for a minute, calming their breathing and heart rate. The rescue swimmer then made a 360-degree scan of the dark horizon for the helicopter. Due to the pitch blackness of the night and high sea swells, it was impossible to see anything. Gibson reached for a MK-124 MOD 0 marine smoke and illumination signal flare from one of the risers on his HBU-11 harness. Using his sense of touch, he felt for the knobs on top indicating night flare versus day smoke and then activated the flare. With his right arm outstretched, he held it high in the air away from the survivor's face. The SH-3D was searching the area for any remnants of life when Howard spotted them. The pilots gave a sigh of deep relief and immediately flew in for extraction.

Earlier when the other rescue swimmer, Williamson, saw the cargo ship coming toward him and the fourth and final survivor, he had pulled them both back into the life raft and grabbed an oar. They both had paddled as hard as possible to barely avoid being crushed by the unyielding vessel. Once clear of danger, he had pulled out his chem light and signaled the crew chief for extraction. They were hoisted up to the helicopter where he secured the victim and then assisted the aircrew's search for Gibson and his survivor. They had major concerns but never lost hope as they hunted for any signs of life.

Gibson remembers everything happening so fast and responding to a situation completely outside of his control. The scenario was so unimaginable that nobody had ever trained for it. He was forced to rationalize a way to push through the impossible. He was extremely exhausted but had to find a reason to survive, while saving another life. He remembers Hell Day during his 1983 Rescue Swimmer School training. Hell Day was a long day of continual pool work and physical training to push the candidates well beyond their perceived limits. After a full day of beat-downs, instructor Bailey continually dropped a heavy brick to the bottom of the pool for Gibson to retrieve. Gibson was exhausted with cramping calves and thighs, but he repeatedly found a reason to dive down and get that stupid brick. No matter what, he found a way to push through the pain and exhaustion to get the job done. This same mentality helped Gibson push through the impossible that night in the Gulf of Mexico. Bill Gibson received a Navy Marine Corps Medal for heroism in a rescue situation that was unique in military history.

Aviation rescue swimmer candidates are constantly reminded of the ethos that they are dedicating their lives to serve others. The coveted U.S. Navy Rescue Swimmer insignia proudly hangs over the entrance to the ARSS training facility. The logo depicts a silver albatross carrying a navy-blue fish in its beak over a golden foul anchor in front of crossed navy-blue swim fins. The albatross was chosen as the rescue symbol from the morbid 1798 Samuel Taylor Coleridge poem, *The Rime of the Ancient Mariner*. The poem depicts an ancient ship stuck in an ice jam in Antarctica, when an albatross appears and leads the desperate crew to safety. The mariner then shoots the rescue bird for food, which angers the ship's crew since it had brought good fortune. As punishment, they tie the dead albatross around the mariner's neck. From that point the boat is cursed. The ship eventually sinks and the crewmembers die, but the mariner is saved by a hermit and his son who believe him to be the devil. Driven by guilt, the mariner spends the rest of his days wandering the earth and telling his story. The albatross symbolizes the selflessness of unconditionally providing hope and safety to others. Not all will recognize or appreciate it, but that's not a deterrent for doing the right thing. Below the SAR insignia in large capital letters reads the simple Rescue Swimmer motto: So Others May Live.

Before training begins, the candidates must read and sign the following volunteer statement, understanding that they are putting their lives at risk to save another. Why would we jump from a perfectly good helicopter in the middle of the ocean in harm's way? Because if we didn't, others would die.

> I AM / AM NOT (CIRCLE ONE) VOLUNTEERING FOR TRAINING AS A HELICOPTER RESCUE SWIMMER AS PART OF MY AIRCREW GUARANTEE PROGRAM. I UNDERSTAND THAT RESCUE SWIMMERS MAY BE REQUIRED TO RISK THEIR LIVES DURING A RESCUE AT SEA. THEY ARE TRAINED TO CONDUCT RESCUES EFFICIENTLY, EFFECTIVELY AND SAFELY. IN MOST RESCUE SITUATIONS, THE SWIMMER LEAVES THE HELICOPTER AND ENTERS THE OCEAN. UPON REACHING THE SURVIVOR, THE SWIMMER REMOVES THE PARACHUTE AND PREPARES FOR HOOKUP TO THE RESCUE DEVICE. A SURVIVOR IN A STATE OF PANIC MAY FORCE THE SWIMMER UNDERWATER, BUT TECHNIQUES TAUGHT TO THE SWIMMER WILL HELP THEM TO OVERCOME THIS RESISTANCE. ONCE THE SWIMMER AND THE SURVIVOR ARE IN THE AIRCRAFT, THE SWIMMER PROVIDES FIRST AID UNTIL MEDICAL ASSISTANCE IS AVAILABLE. TO GRADUATE FROM RESCUE SWIMMER SCHOOL I UNDERSTAND THAT I MUST COMPLETE ALL REQUIREMENTS AS OUTLINED IN THE RSS CURRICULUM.

Aviation rescue swimmers are unique since they tend to operate alone, while in most military jobs you function as part of a team. Leadership and individual courage are hammered into the candidate's training, since once they exit the helicopter they're essentially on their own. The extended team, including the two highly skilled pilots and crew chief, remain 70 feet above, but once in the water it's one-to-one or one-to-many, depending on the situation. They essentially learn hand-to-hand combat in the water, leveraging a series of holds, breaks, and releases to maintain control of the situation. Survivors can be in a panicked state and see the rescue swimmer as flotation, putting them both at risk. The swimmer must assert themselves with confidence to either verbally calm the panicked survivor or get control through brute force and specialized techniques. They call this technique life-saving.

Rescue techniques continually evolve based on lessons learned and real-world data. For many years, the standard life-saving procedures involved taking the survivor deep in the water, which could induce them to panic, then gaining control of them in a controlled cross-chest carry as the rescue swimmer surfaced. Thanks to new Coast Guard rescue swimmer procedures, this was recently changed to more of a "water jiu-jitsu" technique, conducted on the surface of the water. Jiu-jitsu is a Brazilian form of martial arts, based on ground fighting and submission holds. These hand-to-hand combat techniques have been integrated into the existing U.S. Navy aviation rescue swimmer life-saving procedures but occur on the surface of the water rather than underwater. This change was implemented due to the reality that most life-saving battles occur on the surface because of the high buoyancy of the swimmer and survivor's wetsuits, helmets, dry suits, or "Gumby" cold water survival suits, making it difficult to submerge. Plus, forcing a victim underwater and relying on pressure points seems good in concept but, in reality, is unlikely since a victim will naturally release the rescue swimmer once submerged. At that point they are of no use to them since they are panicking for a reason and that reason is the need for air and flotation. Additionally, there are many dangers underwater that should be avoided, if possible, to increase the success of the rescue, while reducing injuries. The value in the legacy life-saving techniques is in the training of learning to remain calm and be meticulous in executing procedures underwater while being attacked. I feel this evolution of training taught me more confidence and to remain calm in chaos, more than any other training I've been through.

In 1993 at the Pensacola Rescue Swimmer School, with eyes tightly shut to simulate complete darkness, I released the side of the training pool and slowly swam to the center in the 15-foot-deep area. As I cautiously side-stroked, I listened for any movement of an active survivor. All I heard was dead silence. I was then suddenly hit hard by the downward force of a member of the previous graduating class acting as a panicked survivor. It felt like I was hit by a train as the wind exited my anxious body.

At the time, the proper procedure in a perfect scenario was to grab a quick bite of air, take three long downward strokes, use pressure-point techniques to break the survivor's hold, and gain total control of them as I swam to the surface with the survivor to then signal the helicopter for extraction. However, training always prepared us for the worst-case scenarios, and I was hit so hard by a panicking simulated survivor that there was no time to get that bite of air. With empty lungs, I forced the survivor to the bottom of the pool where I paused to not only calm myself, but to transfer my body's natural tendency to panic to the survivor. Since I was hit from behind, with his arms tightly locked around my neck, I used a rear head hold release technique. Bending forward to create space between his grip, with both hands I reached up grasping his locked arm. My left hand instinctively located his right wrist while my right hand gripped the middle of his right arm. Lunging forward at the hips, I broke the death grip and moved my right hand close to his elbow and jammed my right thumb into the TH10 pressure point. Applying pressure and dipping my head down and sideways, to not get my mask and snorkel ripped off, I forcefully pushed the survivor's arm up and over my head while twisting it. This effectively turned his body around to where his back was now facing my front. With his right arm now twisted and locked behind his back, I reached over his left shoulder with my left arm and locked him in with a controlled cross-chest carry. That's where both of my arms are wrapped around the survivor diagonally and both hands lock together at the fingers. Through the entire life-saving technique, I maintained the survivor in-close-in-control, which is critical to maintaining control of a panicked survivor. All the while, I kicked to the surface as we both were near passing out from lack of air. As we broke through the surface and took a huge breath of air, I signaled the simulated helicopter for extraction.

Going into a life-saving exercise I was well trained to respond to a hard hit by a panicked survivor. I treaded water with no vision and no idea when the attack was coming but knew how to respond when the hit came. The more important element was my preparedness to not succumb to panic when essentially attacked in the water. Distinguishing, acknowledging, and responding to the signs of increased heart rate and breathing that come from panic are vital to displacing those responses and focusing on the tasks at hand.

We can only plan for things within our control and react to things outside of our control. In this evolution, I knew what was happening so the situation was in my control, but I didn't know when or how, which was outside of my control. Once I was initially hit, I quickly switched my focus to my trained life-saving techniques to gain control of the situation and deliver us both to safety.

* * *

The more you practice self-awareness and controlled breathing techniques and make it a habit the more prepared you'll be to calm your body when it enters fight, flight, or freeze mode. Self-awareness is your ability to recognize the things that make you who you are as a person. It's the mental, physical, emotional, and spiritual state in which the self becomes the emphasis of attention. In this context, self-awareness is more focused on your physical responses (observing your five senses) that often occur once you have a thought that then creates a feeling. Our bodies often start responding without us even being aware of that feeling, and we can jump right into fight, flight, or freeze mode. Thus, the more aware we are of our physical sensations and our feelings, the more control we have over our response.

* * *

In June 2015, the USNS *Mercy* was conducting humanitarian missions in the Indo-Asia-Pacific region near Papua New Guinea. A Helicopter Sea Combat Squadron 21 (HSC-21) MH-60S aircrew consisting of Lieutenant Froslee, HAC and officer in charge; Lieutenant Junior Grade Yaede, co-pilot; and two aviation rescue swimmers, AWS2 Frazelle and AWS3 Courtney, hot-swapped the aircraft with an incoming crew and were planning a day of passenger transfers to nearby islands. (Hot-swapping is when two pilots and two or more aircrew replace the returning crew within a few minutes between fixed-wing aircraft launches.) After strapping into the seats, they received a call from the tower to "Standby, we have a medevac [medical evacuation] for you." The night prior two MH-60S Seahawk helicopters were launched to search for a boat and survivors but came up empty as night approached and both helicopters ran low on fuel. Six souls, including an infant, were in the boat that sank. The young adults swam, carrying the baby, for 3 miles to a remote island. They crawled exhausted up on the shore of Han Island. After some rest, a few of the victims searched the island for any human life. They discovered a small village, which offered a radio to send a distress call. The U.S. Navy responded.

The HAC and co-pilot received a full brief from the tower to ensure they knew what they were flying into. The victims were on an island 80 nautical miles north of their current position. Crew chief and aviation rescue swimmer Taryn Frazelle and her helicopter aircrew reconfigured the cabin of the aircraft for a multi-person rescue versus the previous mission-set of transporting passengers. Frazelle, a new crew chief, and an amazing athlete who had competed in the Regional Crossfit Games, remembers freezing momentarily with an initial spike of anxiety as she received the details of the mission. Aircrew are trained to be transparent in all aspects of the flight to ensure the safety of the crew and aircraft. She immediately communicated her apprehensions to her fellow aircrewman. He responded as anyone of us would in that situation; he yelled at her to focus and snap out of the freeze response. That's all it took. Frazelle pushed her anxiety aside and focused on her extensive training and experience. She went to work, rigging the helicopter for rescue with all the proper equipment, including a rescue litter. She also requested a detachment corpsman, HM1 Hawkins, to help with the medical needs they might require. As they spun up the helicopter, she instructed her rescue swimmer and corpsman to prepare for a mixed overwater and overland medevac.

As they were preparing to launch, they heard from the tower that the SAR mission was called off. The SAR mission didn't have high enough approval. This is common in the field as we tend to hurry up and wait and always be prepared to pivot based on higher command's orders. It's emotionally draining, but a fact of life. The crew began reconfiguring the helicopter for a passenger transfer. And then, just like that, they were told the SAR was approved and to launch as soon as possible. They reconfigured for a SAR mission and launched the MH-60S to the north.

In flight, the HAC spoke with the crew to ensure they were in a good headspace for the rescue since the mission had changed multiple times in the past twenty minutes. It's essential to level-set to have and acknowledge situational awareness of the changes and confirm all of the aircrew attitudes are focused on the mission.

The flight took forty minutes to reach the lone island. They calculated their fuel based on the distance, speed, and new cargo (humans). The only refuel station was the USNS *Mercy*, and there weren't any other SAR assets nearby. They wouldn't have much time on location, so would need to work efficiently and effectively to extract the survivors. When the helicopter arrived on scene, they searched for a safe area but couldn't land due to the island's dense palm tree coverage. The HAC brought the helicopter into a 70-foot hover and Frazelle lowered the corpsman down to assess the survivors. They orbited the MH-60S around the area until the corpsman radioed for assistance. Frazelle lowered the rescue swimmer down.

The helicopter hovered just above the 50-foot swaying palm trees, so the HAC decided to bring it up to 100 feet to avoid any tree-chopping mishaps. Plus, the immense rotor wash can blow debris up into the engines, which would not end well. With Courtney now down on the island assisting Hawkins, Frazelle was the only remaining aircrewman onboard. Typically, the two aircrewmen will man the port and starboard side doors to watch and communicate the aircraft's drift to avoid hazardous obstructions. The crew chief had to watch both sides of the aircraft, crouching and moving back and forth while communicating their position to the pilots.

After receiving a signal from her aircrew on the ground, Frazelle hoisted the survivors up one at a time in a rescue basket and then secured them in the cabin. Hawkins held the eighteen-month-old baby in his arms, protecting her face from debris as they hoisted up together in the basket. Once they were safely onboard, the co-pilot informed the aircrew that they had a transmission oil hot advisory warning light. No secondary indications were present, so they concluded the warning was most likely due to hovering in the hot and humid conditions. There was only one remaining survivor and the rescue swimmer on the sandbar, but for safety concerns the HAC needed to come out of the hover to establish forward flight. He communicated to the aircrewman below via radio to standby as they flew a route to clear the warning light. The advisory light cleared, but being low on fuel they had to be extremely diligent in making one last pass to hoist up the last two. Otherwise, they would need to leave the two on the island while they returned to refuel.

Frazelle calmly hoisted up the Courtney and the remaining survivor, then secured the cabin for forward flight back to the USNS *Mercy*. The three aircrewmen turned their attentions to the victims onboard who were visually terrified, but grateful to be alive. Due to the loud noise of the helicopter and the language barrier, Frazelle relied on hand signals to communicate and to calm the survivors. Forty minutes later the MH-60S landed on the USNS *Mercy* helicopter landing zone. The rescued personnel were escorted down to the ship's infirmary where they were given medical assistance, food, and water.

Aviation rescue swimmers are trained to respond to any kind of rescue scenario. The aircrew attached to the USNS *Mercy* were briefed to fly passenger transfers but then had to abruptly launch to search for missing people on a remote island. Frazelle had a moment of hesitation as she processed the entirety of the mission, which is a normal human reaction to overcoming chaos as we manage our fight, flight, or freeze response. She was able to quickly regain her focus and transition from freeze to fight, which allowed her to effectively save several lives. The crew's success in pulling off the mission with no other SAR assets available came down to good situational awareness, resource management, transparent communication, and understanding and executing their unique roles.

* * *

There are times when we automatically jump into fight, flight, or freeze response, as it can be beneficial in the moment. As a means of short-term survival, it's our bodies' natural way to protect us in a traumatic event as seen in the scenario above.

* * *

In October 2004, Helicopter Anti-Submarine Squadron 15 (HS-15) Red Lions was attached to the USS *John F. Kennedy* strike group operating in the Persian Gulf. Two HH-60H Sikorsky helicopters were working out of the U.S. Army's Camp Arifjan in Kuwait. AWSC Jay Shropshire was a part of the detachment, flying regular sorties into Iraq for CSAR special operations insertion and extractions.

On October 6, Shropshire and team briefed for their mission-set and then spun up the two helicopters. Their mission and crews are redacted from this book since those details are irrelevant for the experience I'm sharing. The HH-60H aircrew were strapped in gunner's belts as they maneuvered in the cabin to prepare for the flight and to man the port and starboard side mounted automatic weapons. The pilots taxied down the runway and made a running takeoff into the wind. The aircraft climbed up and turned left, moving away from the beach until they were feet wet. "Feet wet" is a military term meaning the aircraft is flying over water.

As the aircraft gained altitude and speed, Shropshire heard a strange noise and then noticed hydraulic fluid leaking. He alerted the pilots and crew, and the HAC immediately declared an emergency and pulled the cyclic control to the starboard to return to base. By this time the hydraulic fluid was everywhere, covering the cabin and crew. Fortunately, due to the prompt identification and decision to abort, they safely landed back at Camp Arifjan. They shut down the helicopter and cancelled the mission due to aircraft downed status.

The HH-60H blew a hydraulic line on takeoff, which would have been disastrous if they continued the mission. The highly competent HS-15 maintenance team got right to work to investigate and mitigate the issue to determine if they could get the aircraft operational. Due to the limited number of supplies at the Kuwait detachment, they would need to fabricate a fix to get the helicopter functional to fly out to the aircraft carrier USS *John F. Kennedy* to replace the damaged parts. They worked diligently through the night to resolve the issue, essentially by providing a stint for the hydraulic line and running pressure tests to ensure there weren't any leaks. All tests passed and the aircraft was placed in an operational status for a flight back to the carrier the following day.

The same CSAR crew from the previous day was pre-briefed already, but Shropshire made one significant change for the four aircrew flying in the back. Looking back, he feels it was a divine intervention, but certainly didn't think anything at the time. He made sure the crew was strapped into the cabin seats with five-point harnesses, rather than tethered to the cabin cargo rings via the webbing of the gunner's belts. Belt hoist operation restraints, more commonly referred to as gunner's belts, are military aircraft constraints used by aircrew on military aircraft. They are made of webbing with adjustable straps, which attach to cargo rings placed in various locations in the cabin. The tethering strap is connected to a wider webbing strap that fits around the midsection of the aircrewman and connects via a quick-release assembly. Gunner's belts allow the aircrew to safely move around the cabin of military aircraft to conduct operations, while reducing the risk of falling out open cabin windows and doors. However, for takeoffs and landings aircrew will typically be strapped into the cabin seats since they are designed to absorb strong g-force landings in the rare case of a mishap.

The HH-60H conducted their pre-flight checklist, ensuring the aircraft was operational to fly back to the carrier for full maintenance. The HAC fired up the engines and spun up the rotors. They sat on deck for a fifteen-minute turn to check all gauges. All systems checked green and they collectively felt clear to launch. Taking the same route as the day prior, they lifted and turned left. They were moving away from the beach at an altitude of 250 feet and were almost feet wet when they heard the ungodly sound of metal on metal. Shropshire said it sounded like a deafening screech of dragging metal on the ground. That's when the gauges went red, and all hell broke loose!

The number six drive shaft sheered from the viscous damper, and the helicopter lost all tail rotor effectiveness. The tail rotor pulls against the torque of the main rotor to keep the helicopter straight. In the case of tail rotor loss, the helicopter has no way to counteract the force of the main rotors, causing an uncontrollable high rate of spinning of the aircraft's main body. Pilots and aircrew regularly train for this scenario in controlled environments by practicing autorotations. This is where the pilots will reduce power, drop the collective control (which controls pitch), and step on the right pedal to turn toward and counteract the spin. This aggressive drop in elevation allows gravity to help stabilize the faulty aircraft by using aerodynamic forces rather than the engine.

With a tail rotor loss, it's not gradual, it's immediate. The pilots alerted the crew by announcing "AUTO! AUTO! AUTO!" as they automatically went through their ditching procedures. The crew in the back did the same as they braced for impact while conducting their crash procedures. The HH-60H was spinning uncontrollably at full RPMs, and due to the high g-forces, the pilots couldn't pull back the power control levers to put the engines at idle. They were going down fast and hard, while spinning at horrifying rates. The ICS was cluttered with F bombs as they spun uncontrollably to the ground. The helicopter's crew didn't have time to process the fear before they slammed into the hot desert sand of Kuwait.

The naval combat helicopter impacted the earth at full speed. The main landing gear struts hit so hard they broke through the floor of the aircraft. The violence and sounds of the crash were indescribable as the HH-60H bounced multiple times and then flipped end over end. The crew in complete shock didn't think the violence would ever end as they rolled and rolled. Eventually, the aircraft stopped tumbling and settled upside down.

During the horrific event before the aircraft settled, Shropshire recalls the crew worrying about the helicopter blowing up and wanting to get out of the wreckage. His fight or flight kicked in, and without panic, he took control of the situation. They had all trained feverishly on crash procedures and knew not to exit a helicopter until all motion had stopped. If the rotors are still spinning, then you could easily lose your head, or the aircraft could roll on top of you as you try to escape. He did his best to calm everyone and keep them strapped in until all commotion stopped. One of the door gunners was fervently yelling, "Get out! Get out!" Shropshire yelled, "No! Not until everything stops!" They remained strapped in their cargo troop seats upside down with dust, smoke, and sand covering everything. The main rotors ripped off as they struck the sand, and the main transmission rotor head spun at full RPMs, digging deeper into the earth, until it wound down, getting slower and slower until it ultimately stopped. At that point, they unstrapped and exited to the nine o'clock of the wreckage.

The aircrew conducted medical assessments of each other, and then Shropshire sent in one aircrewman to retrieve the medical kit, rescue litter, and fire extinguisher. There was no movement from the pilots, so the team went in to assess their condition. They were both buried in the sand, and the only way to safely remove them was to dig them out. The HAC suffered from a broken foot, but the rest got away with minor cuts, bruises, and spinal compression injuries. If Shropshire hadn't ensured the crew were flying strapped to their seats, there would certainly have been four casualties from that crash. The gunner's belts would not have held up to the centrifugal force of the crash. He felt a strong calling outside himself to make that change, and his crew is alive today because he listened and responded to that calling.

* * *

We often hear of some people having gut instincts, but truly if anyone is honing in their physical sensations, they might be more aware of times they have a gut response and then choose to act on it. God uses different ways to communicate with us and being mindful is another area of awareness to recognize and listen to how you're being guided. In the example above, Shropshire recognized and listened to his instinct and responded appropriately.

* * *

Since the crash occurred right after takeoff, the ground crew was alerted when the ELT fired off the 121.5 MHz beacon, and they also saw the aircraft go down. The cavalry sped across the desert to reach the crew and secure the crash site. The medical personnel packaged up the two pilots and three aircrewmen, minus Shropshire. He was busy facilitating the casualty collection point when the rest of the crew was taken away. He asked someone nearby whether he should go to medical, and, to their surprise, they thought he was part of the medical response team, not realizing he was in the HH-60H that just went down. He hitched a ride to medical.

Fight or flight kicked in during the egress and recovery of the crash. Shropshire didn't panic but stuck to the methods taught for that dreadful scenario. Many things can go wrong during a flight, and that day they all seemed to happen at once. It's in those moments that we have the opportunity to step up, suppress our fears, and focus on the tasks of survival.

Shropshire served a twenty-three-year Navy career, continually honing the program for future aviation rescue swimmers, and honorably retired in 2014. After retiring he taught Helicopter Underwater Egress Training for Coast Guard rescue swimmers, pilots, and aircrew in Elizabeth City, North Carolina. He was able to share from his real-world experience to help enhance the program and increase the percentage of trainees who completed the program successfully. The most common barrier to completion is the panic that occurs when crashing. And in this case, into water. The students are strapped in blindfolded wearing full flight gear. The simulated helicopter drops and hits the water, then flips upside down. The students must wait for the commotion to stop before calmly egressing out the designated door or window, using nothing but their sense of touch. Shropshire implemented the STOP method to help the students remain calm and focus on survival. STOP is an acronym meaning: S—stop panicking, T—think about what you need to do, O—orient yourself, P—perform or execute what needs to be done to safely egress. The STOP method can be integrated into all aspects of our lives as we find ourselves reaching that point of panic.

As mentioned in chapter one, we tend to live our lives within our safety bubble to ensure we maintain complete control. That may work for a while, but we eventually get thrown for a spin when our lives are disrupted by experiencing job loss, an unhealthy relationship, personal injury, or some other unforeseen life event. We then have an opportunity to respond to our fight, flight, or freeze instinct. If we run from the issue, we'll just continue the cycle and fall into a similar unhealthy pattern. If we have the courage and growth mindset to confront those things outside of our control, then we'll break the cycle of past hurts, fears, and behaviors. We'll retrain ourselves to handle situations in a more confident manner, significantly reducing our anxieties. There's no magic answer to resolving anxiety, but learning and acknowledging areas unique to our personal needs puts us a step closer to a healthier existence.

According to a 2020 survey conducted by SingleCare, 62 percent of respondents experienced some form of anxiety. Low levels of anxiety or fear are actually good, as it keeps us safe in providing a gauge for dangerous situations. Most people experience anxiety to some degree, and aviation rescue swimmers aren't exempt. I personally have experienced generalized anxiety disorder for as long as I can remember, which created a major obstacle for me to overcome. Often my anxiety level was heightened beyond controllable levels, but through continued exposure therapy and learning the proper techniques and responses, I was able to compartmentalize my fears and do my job above the impossibly high military standards. However, that came with a price for me, and many of the rescue swimmers I've had the honor of interviewing. Pushing that distractive fear and pain aside to complete the mission was temporary, as it had to resurface later in the form of physical and emotion release of the trauma like uncontrollable shaking, emotion, being easily startled, and/or post-traumatic stress disorder (PTSD). What we learned was appropriate and effective for saving lives, but it has its longer-term consequences that require attention.

Anxiety is a feeling of fear, dread, and uneasiness. It might cause you to sweat, feel restless and tense, and have a rapid heartbeat. It can be a normal reaction to stress, but there's a point where you need to take back that control before it takes control of you. Special operations units learn a simple habit of box breathing to help remain calm in extremely intense situations. It is often used during fight, flight, or freeze mode to slow down their heart and breathing to re-center and maintain concentration. Box breathing involves four basic steps, which if practiced often will yield immediate results helping you remain calm as well as resting better at night. Think of a four-sided box as your path to breathing. Each side represents four seconds to complete. Calmly breathe in for four seconds, hold for four seconds, slowly exhale for four seconds, hold for four seconds, and repeat. My friend, retired Navy SEAL commander Mark Divine, teaches this method throughout his SEALFIT and Unbeatable Mind fitness programs. He used it during his SEAL training, which helped him graduate as the honor man of his class. I encourage you to search for Mark on the internet for some very informative videos on box breathing and mindfulness wisdom. It can take weeks to months to form a new habit. Box breathing is one of those habits that is well worth the effort, and it's free! You will experience immediate healthy results for your mind, body, and soul. Controlled breathing is the single component that helps feed the brain the necessary oxygen to remain calm in the chaos.

Chapter 5

Situational Awareness

In November 1994, the Kittyhawk-class aircraft carrier USS *Constellation* was operating roughly 100 miles off the coast of San Diego, California. We were heading out on a six-month WESTPAC (Western Pacific) deployment, including time operating in Korea, South China Sea, the Persian Gulf, and Australia. One of our primary missions was to perform exercises off the coasts of Korea after the discovery by U.S. Intelligence that North Korea was attempting to build nuclear weapons. Additionally, we spent three months in the Persian Gulf to patrol United Nations "No-Fly" zones over southern Iraq in support of Operation Southern Watch. The Constellation Carrier Battle Group consisted of an attack submarine, USS *Topeka*; a destroyer, USS *Kinkaid*; two guided missile cruisers, USS *Chosin* and USS *Lake Erie*; an ammunition ship, USS *Kiska*; and a replenishment oiler, USS *Cimarron*. The *Constellation* carried Air Wing Two consisting of ten squadrons: VF-2, VMFA-323, VFA-151, VFA-137, VAW-116, HS-2, VAQ-131, VS-38, VQ-5 DET.D, and VRC-30 DET.2.

It was our first night of flight operations, with the *Constellation*'s full arsenal of aircraft performing night carrier qualifications. Our helicopter squadron, HS-2 Golden Falcons, had two SH-60F helicopters "flying circles" on the starboard side of the carrier. This is commonly referred to as "starboard D" or "plane guard," which puts the SAR aircrewmen in an active ready position to respond to any possible mishap. A quick break in the flight pattern allowed us to fly around the aft of the aircraft carrier and land on the port-side angle deck to conduct an aircrew hot-swap.

The helicopters approach from the aft and creep forward, matching the ship's speed, and then wait for the landing signal enlisted personnel (LSE) to signal and direct the aircraft over for landing. During the day the LSE uses hand signals, but at night they use lighted wands to signal the pilots. The aircraft moves in a slow forward hover toward the landing zone (LZ), halts forward progress over the LZ for stability and accuracy, and then lowers to the deck. The LSE then signals the plane captains to chock and chain the helicopter to the flight deck. This secures the aircraft to the ship while the rotors continue to spin, allowing time to change out the aircrew, refuel, do any minor electrical maintenance, and to rearm ordnance.

As our two Seahawk helicopters maneuvered to the landing area, F-14s and F-18s ferociously idled, with pistons locking the nose gear into the two starboard side catapults waiting for launch. Additional fighter planes anxiously lined up behind the jet blast deflectors, waiting their turn for launch. The jet blast deflector is a safety device, basically a large steel wall, that raises behind the aircraft during takeoff to redirect the high-energy jet engine exhaust and afterburner heat to prevent damage or injury. After the helicopter was safely chocked and chained our shipmates, Justin Tate and John Wandke, hot-swapped the aircraft with me and Dave Hewitt. The flight deck of an aircraft carrier is one of the most chaotic environments imaginable. The ear-shattering noise of the aircrafts, catapult launches, and arresting wire recoveries leaves all personnel with permanent hearing loss, even with the use of hearing protection. Dave and I were on the ICS providing a turnover debrief to Justin and John, while the pilots did the same up front on a separate ICS channel. Turnover debriefs inform the onboarding crew of any aircraft discrepancies, rescue devices used or needing replenishment, and any pertinent information that affects the success and safety of the next flight.

Dave and I unplugged our ICS cords, stowed them securely in our flight bags, and motioned for the LSE's attention. The common smell of JP-5 or Jet Propellant 5 fuel overwhelmed my senses as I looked around, taking in the precision actions of the military professionals on the pitch-black flight deck. Standing in front of the SH-60F helicopter, just outside its violently spinning rotors, the LSE acknowledged us and waved his lit wands toward our starboard side, outside the rotors, indicating it was safe for us to exit. Carrying our heavy, green personal flight and rescue gear bags, we lowered our helmets and walked directly out and then around the LSE to safely remain outside the spinning rotors. We then disappeared into the darkness down the carrier's port side catwalk and into our parachute rigger shop to store our gear in our individual cages, before debriefing with our aircrew.

As helicopter Hunter 616 took off to reestablish the "starboard D" pattern, the fixed-wing aircraft launches and recoveries commenced. Hunter 616 flew a few miles ahead and off to the right of the carrier to conduct nighttime sonar dips for anti-submarine qualifications. The SH-60F was equipped with a variable depth dipping sonar and sonobuoys to detect and track enemy submarines. It can also carry MK-46, MK-50, and MK-54 torpedoes to neutralize enemy subs. Tate, strapped into the port side acoustic sensor operator seat, lowered the sonar dome to 500 feet to check the water temperature, sound variances, and fired a couple active pings. Fun fact: A few months earlier, Justin Tate actually hooked the sonar dome directly on a submerged U.S. Navy submarine's main sail, which terrifyingly dragged the helicopter a few feet before it severed the dome. The odds of that happening are one in a million.

As Hunter 616 was raising the sonar dome, the crew received a radio call for a man overboard. Tate quickly completed the sonar operation by raising and seating the dome, unstrapped from his seat, and began donning his rescue swimmer gear. Wandke rigged the cabin for rescue, getting the rescue strop ready with a lit chemical light attached. The pilots hastily flew the aircraft back toward the ship to search for the survivor in the water. That's when another radio call came through, clarifying that it was a "pilot" man overboard. At this time, there was still no indication that an airplane had crashed, or a pilot ejected. This is very common during an emergency when misinformation is initially provided, but then updated later with new accurate data. Aircrew are trained to be ready to respond to any issue, knowing the mission-set they launch for often changes on a dime.

The helicopter pilots soon spotted the bright, white reflective helmet of the survivor bobbing in the high seas. They flew one pass over the survivor where the crew chief, Wandke, marked his position by tossing out a MK-25 flare from the cabin door. The MK-25 is a saltwater-activated smoke and flame flare used for marking survivors in the water. The crew chief assessed the area for hazards that could cause an issue for the rescue and communicated to Tate what to expect. At this time, it appeared more likely that someone fell overboard. If it was a fixed-wing ejection, there would be a life raft and parachute in the vicinity, which was not visible from the helicopter. Hunter 616 then flew a nighttime wind-line SAR pattern approach toward the illuminated flare and hovered slightly left of the survivor to deploy the rescue swimmer. The crew chief used the aft hover trim controls to position the helicopter over the survivor before lowering Tate. The aft hover trim control is a mounted joystick with a radio and ICS button, hoist controls, and ability to move the helicopter when in a hover after authority is transferred from the pilots. Since the crew chief has eyes on the survivor and rescue, it helps to provide that control so they can direct the aircraft in an efficient and effective manner to expedite the rescue.

Some clarity on the mishap that led to the survivor in the water. An F/A-18C Hornet pilot came in for a night aircraft carrier landing, but his tailhook missed the arresting cables. This is a common scenario, where he properly followed the standard operating procedures to ignite the afterburner to gain the appropriate speed to take off and then fly around for another landing attempt. However, due to a malfunction, the afterburner neglected to ignite, and the F/A-18C rolled over the angle deck and dropped straight down toward the ocean. The pilot had milliseconds to react as he grabbed the ejection handles and punched out. He safely ejected from the aircraft, but because he was so low to the water, his parachute didn't have time to fully deploy. As the fighter pilot entered the frigid water, the USS *Constellation* immediately paused the catapult launches and arrest landings, then turned the 1,088-foot ship to the starboard to avoid crushing the stranded pilot. As the pilot descended to the ocean, his crashing F/A-18C's afterburner automatically ignited after a short delay. The ghost aircraft, missing a canopy and pilot, roared up into the night sky with a streak of trailing fire. The naval officers in the aircraft carrier tower scrambled with ideas to send another jet to shoot down the unmanned Hornet. However, that wasn't necessary as the F/A-18C flew an arc up and over the aircraft carrier's superstructure, splashing into the pitch-black Pacific.

The Hunter 616 helicopter pilots held a 70-foot hover into the wind, while the crew chief moved the aircraft near the survivor and then lowered the rescue swimmer down the hoist. During nighttime rescues, the helicopter uses a spotlight while the rescue swimmer has a lit chemical light attached to the top of their mask. In complete darkness the small chemical light illuminates the area fairly well, but the spotlight can quickly reduce your eye's dark adaptation. The rescue swimmers are trained to identify and respond to all aspects of the physiological components that can affect the outcome of an operation. It takes twenty to thirty minutes for the rod cells in the human eye to adjust to night vision, but only seconds to adjust back to light. A trick when exposed to light is to close one eye until the light goes away. Since eyes are coordinated, the brain adjusts to equalize the dark adaptation. It's said that pirates wore patches over one eye exactly for this purpose, to retain their ability to see in the dark. By closing one eye on the descent, Tate was able to maintain half of his night vision to increase his effectiveness during the night rescue.

As Tate got closer to the water, the strong kerosene smell of JP-5 jet fuel was present, which is not indicative of a man-overboard scenario. Thinking he might have incorrect intel on the mishap, he scanned the area toward the survivor to see if a parachute was present. He couldn't see anything other than a person rising and falling with the high sea state. As he continued being lowered and approached the ocean, Tate kicked his rocket fins to displace some of the static discharge created from the helicopter's spinning main rotors. When safely grounded out in the water, he detached himself from the rescue hook. Tate raised his arm with an open palm facing upward to signal "OK" to the crew chief and then turned and swam 20 yards toward the survivor bobbing in the sea. That's when Tate realized this was an actual pilot in the water, which could still mean a man overboard since pilots can be blown off the ship when working on the flight deck. The F/A-18C pilot was safely floating, but there still wasn't a parachute visible and Tate noticed a more pronounced smell of jet fuel in the water.

The SH-60F crew chief shifted the hovering aircraft left and back to provide space for the rescue swimmer to conduct the rescue. Tate established communication with the survivor over the high-pitched helicopter twin turboshaft engines and thumping rotors. He then grabbed ahold of the pilot's flotation from behind and started kicking his fins hard to get moving toward the distant rotor wash. That's when he kicked the Hornet pilot's seat pan. Military aircraft equipped with ejection seats house a seat pan that contains survival equipment, including a life raft, distilled water, batteries, radios, and other essential gear. The seat pan typically releases when immersed in saltwater, but his was still attached to the pilot's harness either due to a malfunction or possibly the low-altitude ejection. As Tate towed the pilot through the water, the motion activated the partial release of the pan and then the life raft popped out and automatically inflated. This changed the scenario because seat pans are only attached to a pilot if they ejected, which validated this wasn't a standard man overboard as the initial radio call suggested.

Where is the parachute? kept going through the rescue swimmer's mind. He meticulously went through his rescue procedures to detach the pilot's seat pan from his harness and then made sure he was clear of any unnecessary gear that could get hung up on either of them and drag them under. However, since the parachute never fully canopied from the ejection and from the dark, high sea state, the parachute remained hidden just under the surface of the water. While conducting his above and below surface checks in the darkness, Tate found the saturated parachute after he swam into the middle of it. Like a fast-acting spider cocooning its prey, the chute wrapped the web of shroud lines around both the rescue swimmer and the pilot.

Rescue swimmers train immensely for this very scenario, and with low visibility and high seas, Tate remained calm and quickly assessed the rapidly changing situation. He narrowed his focus and communicated the immediate tasks required to get them both to safety. Tate remembered his tangled parachute training from Aviation Rescue Swimmer School. First, determine the wind direction, which can be done by looking at the hovering helicopter. The helicopter will always hover into the wind. Second, slowly and methodically swim into the wind with small hand and flipper movements. This will work with the current and wash the parachute off you. As he started these procedures, he could feel the parachute releasing and starting to wash away. Tate systematically removed the parachute and shroud lines from himself and the pilot, while never losing physical contact with the survivor. One by one, he pulled and pushed the tangled lines, knowing he could use his j-hook knife as a last resort. However, when you cut a line, one becomes two, which is why it's a last resort, as it can create more harm than good. He released each line while simultaneously towing the pilot into the strong current and helicopter-created gale force winds. Once he and the pilot were clear, he tightened his cross-chest carry and started kicking to put as much distance between them and the parachute as he could. Tate then performed one final check of the survivor and signaled to the Hunter 616 crew chief for extraction.

Wandke continuously communicated the rescue operation to the pilots, who relayed the information to the aircraft carrier tower. Another HS-2 helicopter was hovering nearby with a rescue swimmer dressed out in case they needed assistance. Fortunately, Tate was able to adapt to the changing intel and environment and flawlessly perform the rescue. The crew chief used the crew hover trim to move in for extraction and both were quickly hoisted to the helicopter. Wandke performed initial medical checks of the pilot as Hunter 616 landed back on the USS *Constellation*, where the pilot was taken to the ship's medical for a more in-depth assessment. The aircrew was awarded the Navy and Marine Corps Achievement Medal for their quick action and professionalism in saving the F/A-18C's pilot.

No matter the scenario in life, assessing the situation allows you to get a full grasp of all the elements involved. Aviation rescue swimmers are meticulously taught this basic skill to ensure they enter a rescue with as much information as possible to help increase their odds of success. However, it's not commonly possible to know every detail until you're immersed in the rescue. As Justin Tate found with his initial misinformation and a hidden, saturated parachute, he had to quickly reassess the situation and respond to effectively complete the rescue.

The first action rescue swimmers are trained to do in all search and rescue scenarios isn't a physical action, but an analytical action—to assess the situation. Look. Listen. Learn. They need to ensure they know what they're facing to map out a plan of attack. If they blindly jump into a problem, they may get lucky, but the more likely scenario is they won't be prepared and will cause unnecessary work and risks. In most cases, the rescue swimmer will either already be airborne or launching with some intelligence relayed from an emergency response dispatcher. This could literally be any scenario you could possibly imagine. Anything from a sailor walking off the front of the aircraft carrier flight deck, which did happen on our 1995 deployment, to a commercial jet airliner going down off the coast of Mexico, which also happened during another deployment. Based on limited and developing intel, they launch with the appropriate rescue gear, rescue personnel, and support aircraft, which makes them a SAR asset. Additionally, the ship will alter its course toward the disaster to close the gap and call-in additional SAR assets as required.

The rescue swimmer candidates prepare for worst-case situations within reason and within the safety boundaries of their training regimen. There are always situations that exceed what our minds can imagine, so it's impossible to cover all potential scenarios. The goal isn't to simulate every imaginable situation, but to create difficult scenarios that help give the students the confidence and skills required to perform in the real world. Each student conducts multiple survivor rescue scenarios, which they call "multis." The multis differ from student to student and increase in intensity as they progress through the program. Looking back, I realize now how fun the multis were, but in the moment when my career depended on my ability to pass them, it was extremely stressful.

New York Times bestselling author Malcom Gladwell describes the adaptive unconscious in his 2005 book, *Blink—The Power of Thinking Without Thinking*. Adaptive unconscious is the ability to "thin-slice" a limited amount of information to come to a conclusion. Aviation rescue swimmers train in difficult and chaotic situations to become accustomed to making decisions on limited information and readjusting as the scenarios change. The more they train and are exposed to chaos, the better honed their adaptive unconscious evolves. This is a normal philosophy with military and first responders. And the opposite holds true as they get further away from the training and traumatic environments, the slower their responses will become as their brains are forced to filter through the information. The ability still remains, but the response times will decrease as they distance themselves from the fight.

The candidates are all huddled in the locker room shower area dressed in full SAR gear consisting of a short-sleeve wetsuit, deflated LPU-28P / SAR-1 flotation vest, and HBU-11/P rescue harness, while holding in one hand their fins, mask, and snorkel. The rescue harness contains two MK-13 / MK-124 MOD 0 flares and smokes, j-hook pocket shroud cutter, SAR Scabbard knife, two high-intensity chemical lights, two general purpose chemical lights, strobe light with blue detachable lens, lifting strap with V-ring and snap hook, and an AN/PRC-125 radio. These older rescue harnesses were made of webbing with shoulder risers and straps that fed around the swimmer's back and securely connected in the front. They didn't provide any lower body support, so the load was transferred to the rescue swimmer's back when hoisting 70 feet up to the helicopter, which has attributed to long-term back issues for most SAR veterans. Years later, the harnesses were replaced with the TRI-SAR rescue swimmer harness. The TRI-SAR is a helicopter hoistable full-body rescue harness with integrated flotation. With leg loops added to the harness, the weight distribution, comfort, and effectiveness increased tenfold, plus hopefully reduced long-term back issues.

The students sit in anticipation, not knowing what to expect beyond the locker doors. As soon as their individual name is called, they grab their mask and fins and hustle to the training pool. The instructors immediately create a situation of distraction and chaos, which the students must quickly adapt to. Classes used to come up with a chant they memorized and recited while being screamed at by an instructor and sprayed down with cold water from a hose. My class chant was written by SAR student Keith Redmond, who would flawlessly make it through the pipeline and serve with HS-8, followed by a long career in law enforcement in Louisiana. It went, "Helos, Jets and Turbo Props. We are the class that never stops. Saving lives is our job, souls from the reaper we will rob. A 4.0 sailor is the key. That's what we are: air, land, and sea!" We were all jacked high with anxiety and amped up to save lives, so much that the simplest task tripped a lot of us up. We had to slow down and overcome a mental obstacle before moving onto the pandemonium in the helicopter tower and training pool.

The multi setup and scenario change based on each new candidate, so I'll describe one possibility. The student climbs the tower, which is a simulated helicopter suspended over the deep end of the pool equipped with a rescue hoist, high-powered sprinklers emulating a small example of the intensity of rotor wash, and a spotlight for night rescues. The only thing missing is the high-pitched screaming of the engines and rotors, plus the hurricane force of actual rotor wash. The screaming instructors and survivors do their best to make up for the lack of sound, but nothing compares to the deafening noise of operating under a helicopter. In the cabin, a crew chief frantically barks information about the desperate survivors in the water. Despite the magnitude of chaos, the student must focus on personal safety first. They connect a gunner's belt around their chest so they don't fall out of the helicopter. An instructor won't hesitate to fail the "dead" candidate if they miss that vital detail.

The student quickly dons their fins, places their mask with snorkel positioned on their forehead, and slides to the cabin door and swings their legs out. Sitting in the cabin is the first opportunity for the candidate to visually assess the situation. They will take into account the sea-state, debris and fuel in the water, the number of survivors, and their distance and critical state. With a screaming crew chief instructor, the rescue swimmer must find their calm as they scan below to see if what they see aligns with the intelligence that was shared by the instructors. It's highly unlikely that it will.

With a sense of urgency, the crew chief smacks the student's chest on top of the quick-release mechanism in the center of their gunner's belt. This is their indication they are approaching the jump speed and altitude of 10 feet and 10 knots. They release the belt as the crew chief yanks it away to ensure the rescue swimmer doesn't get snagged while deploying from the helicopter. This is followed by three consecutive taps on the student's right shoulder, which lets them know it's time to jump. There's a good chance they will try to trip the candidate up by only tapping twice, in which case they better not jump. Each test or trick may be a disqualifier, but they won't know until they've completed the entire rescue. After three confirmed taps, the student looks left, looks right, and through their legs for any debris before pushing off the cabin floor. It's at this time they let gravity take over.

Jumping from a helicopter may look graceful to an outsider, but those that know, know. Rescue swimmers are fully equipped with rescue gear and need to be diligent about their water entry or risk creating an ocean yard sale of gear or getting injured. The rescue harness has a lifting V-ring attached to webbing, which if left unattended risks hitting the swimmer in the face and knocking their teeth out, which isn't good. The mask and snorkel rest on their forehead, but on entry it will rip off from the down force of the water. They tether it with a string attached to their harness just in case they need to retrieve it. Jumping with fins adds another complexity since they don't want to land on their tips, as it'll slam them face first. Wherever the rescue swimmer looks is the way their body will land. If they look directly down at the ocean, then they're going to flip uncontrollably toward the water. It doesn't help the situation if the rescuer gets injured or loses gear on their deployment. It just adds one more victim to the problem.

To combat all these variables, the rescue swimmer pushes off from the cabin floor, never losing eye contact with the horizon. This puts them in the water at a perfect L-shape, with fin tips pointed up to enter on the heels of their fins. They cross their right arm across their chest, with their right hand placed on their left shoulder pinning down the harness-lifting V-ring to prevent it from knocking out their teeth. With their left elbow bent across the outside of their right arm and forearm straight up, they press their left hand on the front of their mask suctioned to their forehead. This procedure is practiced over and over to ensure it becomes automatic. Rescue swimmers immediately clear their mask to reassess the situation from underwater. Upon water entry they slide their mask down to its proper location covering their eyes. They press the top firmly to their forehead and force exhale from their nose. The air displaces the water from the flooded mask. At that point, they have as clear vision as possible below the water, which isn't very clear, but it makes the rescue swimmer operable once they surface.

In a multi training evolution, there are typically three survivors in the water and additional victims requiring advanced first aid in the helicopter. Of the three in the water, as assessed from the helicopter, number one is an F-14 pilot still attached to her parachute draped over her and shroud lines tangled all over her body. She most likely has her oxygen tube and survival seat pan still attached to her harness, all of which are major risks of suffocation and drowning. Survivor two lays in a single-person life raft, but the student can't detect any movement, which could be an indication of a cardiac arrest or unconsciousness. Survivor three is an active and panicked survivor, and from underwater the student sees him searching for the attack as soon as the rescue swimmer surfaces.

Survivors one and two are the obvious priorities to get to safety, but survivor three isn't going to allow that. The student dives deep below the active survivor and surfaces behind him, immediately locking him in a controlled cross-chest carry. The free-floating survivor thrashes and rolls, trying to get loose as the rescue swimmer candidate shouts "words of encouragement" to calm him down. The student must release one hand to signal the helicopter to lower the hoist for extraction. Survivor three takes advantage of the situation and spins to lock his arms around the swimmer. Without hesitation, the rescue swimmer performs something they call suck, tuck, and duck. That's where they grab a quick bite of air, tuck their chin, and duck underwater. The student takes three swift arm strokes down and then conducts life-saving procedures to break the survivor's hold, using pressure points and swift movements to turn him around, and places him back in a controlled cross-chest carry. This technique has since been replaced with the updated "water jiu jitsu" life-saving techniques of remaining on the surface to gain control of the active survivor. With one of his arms locked behind his back to encourage his compliance, the student resurfaces the constrained survivor and signals for helicopter extraction.

Soon after they surface, the rescue swimmer looks up and notices the rescue hook being strategically lowered on top of them and quickly pulls them both back underwater, clear of the hook. The helicopter main rotors generate a tremendous amount of static electricity as they spin, which travels down the cable and can electrocute the swimmer and survivor if they touch it before allowing it to ground out in the water. The rescue swimmer resurfaces and wraps the now calm survivor in the rescue strop, attaches the safety strap around his chest, and gives a thumbs up to the crew chief. As survivor three is hoisted up to the helicopter tower, the student turns back to the remaining survivors.

The two remaining survivors bring additional challenges, as nothing comes easy. Survivor one needs immediate attention given that she is being dragged underwater by her parachute and is suffocating from her attached oxygen mask and hose submerged in the water. Her objective hazards must be removed, and then a timely rescue litter extraction is needed due to a suspected spinal injury. Survivor two floats safely in a life raft but suddenly goes into cardiac arrest. In the end, the rescue swimmer may not be able to save them all, but they're sure as hell going to try! There isn't always a perfect solution, but it's how you reach the decisions and learn from them that matter when being evaluated.

Humans have a natural response of fight, flight, or freeze when we encounter fear and danger. The rescue swimmer instructors make sure students demonstrate the capability of an automatic response to both fight and flight, based on the dangerous scenarios they put them through. There is absolutely no room for freezing! Losing composure and becoming paralyzed puts all the lives at risk, including the rescue assets.

In June 6, 2014, the USS *Bataan* and USS *Elrod* answered a distress call sent in from an Italian military marine patrol aircraft. They reported seeing six small vessels crowded with refugees in the Mediterranean Sea, one of which was sinking. Two MH-60S helicopters from Helicopter Sea Combat Squadron 22 (HSC-22) launched from the *Bataan* as SAR assets. The *Bataan* deployed a 7-meter rigged hull inflatable boat (RHIB) and one 11-meter captain's gig, and the *Elrod* deployed their 7-meter RHIB, all containing the ships' surface rescue swimmers.

Lieutenant Commander Addison Daniel and Lieutenant Jerrid Stottlemyre piloted one of the aircraft with Crew Chief AWS2 Brandon Coan, AWSC Wade Hove, and AWS2 Kevin Gordon in the back. The aircrew was quick to launch when they got the call and were on scene within minutes. Aviation Rescue Swimmer Gordon prepared for water entry as Coan and Hove rigged the cabin for rescue. The pilots brought the MH-60S into a wind-line SAR pattern and Coan tossed out a pre-armed MK-58 MOD 1 smoke. This was when the rescue crew first established visual contact with the six life rafts, all filled with desperate African refugees and some already in the water hanging onto the sides of the rafts. Gordon assessed the situation from the cabin door as he planned out his rescue priorities. He knew this was going to be a big one!

The HAC brought the MH-60S into a slow hover as Coan prepared to lower Gordon near the sinking raft. The extreme force of the rotor wash pushed the rafts away making the refugees duck and scream as they tried to protect their exposed faces from the spray. The rescue swimmer entered the water, disconnected from the rescue hook, gave an OK hand signal to the crew chief, and then set his sights on the most critical looking life raft. He moved quickly through the low sea-state with the rotor wash working to his advantage, pushing the current against his back.

Gordon established communication with those treading in the water near the sinking raft. The refugees urgently pointed to a lifeless man requiring immediate medical assistance. The man had blood coming from his eyes, ears, and nose and had been burned both from fuel in the water and the scorching tropical sun. The rescue swimmer attempted to make physical contact with the unresponsive refugee but quickly stopped when he felt the skin peeling from the victim's limp hand. Gordon reassessed the unresponsive man's needs, and necessary caution based on this new information. He decided to gently roll the man out of the raft and into the cool water to reduce further injury.

As soon as the man submerged in the salty water, he instantly became active and viciously attacked Gordon. His frail and pruned hands reached out and ripped the rescue swimmer's mask off and locked his legs around him, trying to use him as flotation. Like muscle memory, Gordon performed a life-saving technique to break his hold, but the panicked man was clearly trying to kill him. Without hesitation, Gordon struck the man in the face, stunning him for a brief moment as he quickly regained dominance of him in a controlled cross-chest carry. A nearby surface rescue swimmer had just arrived on the captain's gig and deployed into the water to assist. Between the two of them they were able to safely get the man onto the rescue boat. The USS *Bataan*'s medical staff would later determine the survivor was suffering from the "Triad of Death," hours from his own demise. (From Wikipedia: "The trauma triad of death is a medical term describing the combination of hypothermia, acidosis, and coagulopathy. This combination is commonly seen in patients who have sustained severe traumatic injuries and results in a significant rise in the mortality rate. Commonly, when someone presents with these signs, damage control surgery is employed to reverse the effects.")

Gordon snatched a rescue flotation device from the captain's gig to assist with shuttling survivors to the boat. He worked with the other rescue swimmers hauling refugees one by one from the life raft to the gig or RHIB. After thirty minutes of continuous swimming, Gordon had brought a total of eleven survivors to safety. After validating all the refugees were safely on the gig, he returned to the empty raft where he unsheathed his SAR knife. He stabbed the side of the raft and made sure to scuttle it. He then gave a thumbs up to his crew chief and they moved in for extraction. Amped and dehydrated from the multi rescue, Gordon sat on the shaky cabin floor and rehydrated and calmed his breathing as he prepared to go back in.

The pilots repositioned the MH-60S helicopter near another life raft containing seventy additional refugees both in and out, holding onto the sides for flotation. Gordon attached himself to the rescue hook and Coan lowered him back down. Halfway down the hoist he noticed the strong helicopter rotor wash pushing the raft a hundred yards away from their position. Frustrated, Gordon entered the water and detached from the hook and began swimming the extra distance. As he approached the raft, he removed his snorkel from his mouth and established communication with the crowd of desperate victims. After catching his breath, he shouted, "I'm a U.S. Navy rescue swimmer and I'm here to help!" Gordon recalls the joyful cries from the refugees as they started chanting, "New York! New York! New York!" Gordon later realized that the refugees probably only heard him say "U.S." and thought they had floated to New

York. They were ecstatic since they thought they were being brought to the United States.

The rescue swimmer was fortunate to find an English-speaking man who helped translate his rescue intentions, to individually tow each person to the helicopter to be hoisted up. One man entered the water and Gordon hauled him toward the rotor wash as the pilots moved closer to reduce the distance. It was a balance of closing the gap, while preventing the uncontrolled forced drift of the raft. They arrived under the helicopter and the crew chief lowered the rescue basket. Gordon placed the victim securely in and gave a thumbs up to Coan. The basket cleared the water, but then halfway up it suddenly jolted to a stop causing a shock load to the unknowing survivor. There was a hoist malfunction preventing the reel to safely lift and the crew chief had to quickly lower the confused refugee back into the ocean. Coan used hand signals to inform Gordon of the emergency. Gordon acknowledged and then attempted to communicate the situation through the intense rotor wash and noise to the puzzled refugee. The rescue helicopter needed to return to the ship to be repaired, leaving Gordon and his survivor now 200 yards from the raft. Rather than waste unnecessary time complaining, the rescue swimmer grabbed his victim and headed back to the raft. He arrived at the life raft and climbed onboard with the exhausted and wet refugee. From there, Gordon used his radio to provide real-time updates to the helicopter and boat SAR assets to assist with on scene coordination.

The U.S. Navy rescued 282 survivors that day. Gordon personally rescued eleven, including a combative man suffering from the "Triad of Death" and assisted with the logistics and coordination of rescuing several others from the life raft. Despite the many challenges and resistance he encountered, Gordon remained calm and focused. He recalls a lot of things that didn't occur by the book, but failure was not an option and he adapted to the various obstacles. He was awarded the Navy Marine Corps Commendation Medal for heroism. His entire MH-60S crew would later receive the 2014 Aircrew of the Year award from the Naval Helicopter Association.

* * *

When life doesn't go as planned, as it often doesn't, how do we adapt to those unexpected trials? We can push in and persevere or pull away and withdraw. Success is often the result of our hard work and our ability to keep pushing through, especially when it seems too hard, or we experience failures.

* * *

On August 2, 1989, Helicopter Anti-Submarine Squadron 14 (HS-14) participated in one of the largest sea rescue operations by a helicopter squadron in recent naval aviation history. They were returning from a six-month WESTPAC deployment on the USS *Ranger* and about to begin flight operations. The warship was 500 miles east of Cam Rahn Bay, Vietnam, steaming through thunderstorms, which were affecting the aircraft carrier's flight operation schedule. A Navy A-6E Intruder from attack squadron VA-145, flying a routine mission, spotted what looked like a fishing vessel adrift near the USS *Ranger*. They informed the carrier, who diverted an HS-14 helicopter from plane guard to investigate.

Two HS-14 Chargers aircraft were airborne at the time of the investigation request. One SH-3H Sea King was flying plane guard and the second was flying a battlegroup logistics mission. Lightning 611 was the first helicopter on scene, piloted by executive officer Commander Dave Dahmen, Lieutenant Junior Grade John Begley, Crew Chief AW1 Mark Klausmeier, and Aviation Rescue Swimmer AWAN Felix Wyatt. Once vectored to the area, they were able to identify the vessel in distress and report back. It appeared to be a foreign barge with about twelve male victims on the deck. It had a large Quonset hut in the center with two 30-foot masts at the fore and aft of the barge connected by cables. This made a vessel hoisting rescue option too dangerous, which meant the victims would need to enter the frigid ocean water. Additionally, the barge was swamped by the high seas. It was a matter of time before it sank, increasing the urgency to get the victims to safety.

The HAC brought the Sea King into a SAR pattern while the crew chief armed and tossed out a MK-58 MOD 1 smoke. The SH-3H flew an extended circle and made its final approach into the wind. Wyatt sat in the doorway, dressed in his full wetsuit and SAR gear, legs and fins dangling, amped and ready to do what he was trained for. The helicopter approached 10 feet and 10 knots, kicking up a cyclone of rotor wash. The HAC communicated to the crew chief, "JUMP! JUMP! JUMP!" Wyatt looked below for any debris in the water, then pushed off the metal floor and entered the freezing South China Sea. As his adrenaline met and neutralized the shock of cold water, he surfaced with an OK hand signal, then turned to swim toward the barge.

High waves, strong currents, and gale winds created a challenge for his 20-yard swim to the sinking corner of the barge. Once he made contact, he climbed up on the rickety platform, removed his fins, and assessed the situation. None of the refugees spoke English and he assumed nobody had any water survival training. Wyatt used hand gestures to explain how the rescue operation would be conducted and that each person would need to be in the water to be hoisted. The young rescue swimmer wasn't sure how much of his explanation was landing or not, but he figured he'd better demonstrate by rescuing the first lucky volunteer. Wyatt donned his fins and reentered the water. A volunteer refugee jumped in next to him. Wyatt then placed him in a cross-chest carry and towed him to a safe area clear of the barge and signaled for Klausmeier to lower the hoist. One by one, the rescue swimmer towed the survivors from the barge to a safe distance to be hoisted up. This evolution was time consuming and exhausting work, but he relentlessly continued saving lives. After Wyatt hoisted up with his seventh survivor, a second helicopter took over as Lightning 611 returned to the carrier.

The fierce sea-state 5 (8-foot waves), high wind, and torrential rain made for extreme conditions to rescue the stranded survivors. The USS *Ranger* launched most of the HS-14 helicopters to assist with the rescue. A couple SH-3Hs remained on the carrier with Ready Alert aircrews dressed and standing by to launch if needed. The HS-14 maintenance crews remained on deck in the poor weather conditions, working around the clock to ensure each returning helicopter was fueled, restocked with ordnance, and fixed if any minor mechanical or electrical problems needed addressing.

Lieutenant Commander Mike Fackrell, Lieutenant Jeff Maclay, Crew Chief AW2 Phil Gonzales, and Aviation Rescue Swimmer AW2 Jeff Bast were next on scene. The SH-3H HAC made their SAR approach and deployed the rescue swimmer. Bast ran into immediate challenges as two refugees jumped in the water before he could reach the barge. He struggled to communicate with them due to the language barrier and their panicked state. However, he maintained operational authority and was able to gain control of each victim and tow them to a safe location to be hoisted. At this point, Bast noticed several small sharks circling the vicinity, but fortunately they didn't interfere with the operation. When Bast signaled Gonzales to lower the hoist, the pilots moved in for extraction. With that brought the intense rotor wash, which elevated the victims' panic to another level. Bast tried communicating with hand signals, but it was pure chaos in the center of the helicopter's hurricane. The victims kept trying to grab the cable instead of the horse collar. Bast was forced to aggressively gain control of the situation and secure them into the hoisting device against their retaliations. It was strenuous work, but he was able to save five of the victims before he was hoisted up and returned to the USS *Ranger*.

While Lightning 611 was conducting an aircrew swap on the aircraft carrier, the oncoming aircrew noticed a massive squall approaching. They called the tower to request an S-3 Viking from VS-38 to mark the barge's location with a smoke and sonobuoy. Visibility was less than one-half mile, making it extremely difficult to have multiple aircraft working closely together during the rescue. They were utilizing numerous resources from the carrier air wing to coordinate and safely execute the operation. The oncoming crew was Lieutenant Bob Ernst, Lieutenant Aaron Flannery, Crew Chief AW1 Mike Walsh, and Aviation Rescue Swimmer AW3 Charles Harcus.

Harcus entered the water near the sinking vessel, but due to dangerous swells, decided against climbing up on the barge. Instead, he remained in the water and instructed victims to jump in one by one to be rescued. He immediately noticed the heavy film and pungent smell of fuel on the surface of the ocean. The rescue swimmer fought through the rough water, doing all he could to keep his victims' heads above the fuel. By the time Harcus towed and signaled Walsh to hoist up two survivors, the other helicopter joined back into the mix. The other SH-3H had just returned from offloading their five survivors and replaced Rescue

Swimmer Bast with AW3 Randall Fischer. While Harcus continued rescuing refugees, Fischer deployed from his helicopter and swam 20 yards to the barge. He was able to climb on board and assess the remaining victims and their medical needs. The rescue swimmer identified five children and several more adult males and females. He relayed the information to the hovering helicopters via his handheld radio. The helicopter, commanded by Lieutenant Ernst, moved in for extraction of the children since it was equipped with a Billy Pugh rescue basket. The basket is a much safer enclosure for lifting babies or small children, since they don't have the upper body mass or strength to fit securely in a rescue strop.

Fischer pulled a two-year-old toddler into the water and began towing the fragile victim through the high seas toward a safe extraction area. While performing strong flutter kicks to distance himself from the barge, Fischer snagged a submersed mooring line, which tore off his fin. The rescue swimmer held the crying child above the water as he kicked his long-stretched legs for propulsion while submerged in the fuel-saturated ocean. With the high sea-state and wind, having a single fin and no usable arms to pull, it was like treading water without making any forward progress. The work to just keep the child suspended and protected took every ounce of energy, but he never quit. Fischer finally understood why it was so important to demonstrate holding bricks above his head in the pool during training.

Harcus noticed from a distance that Fischer was struggling and diverted himself and the survivor he was towing toward his peer rescue swimmer. After establishing contact, he assisted Fischer in getting the child safely to the rescue basket and then hoisted up with his adult survivor and the toddler by Crew Chief Walsh. Harcus dislocated both of his shoulders as he held tight to the thrashing basket on the final hoist, ensuring both survivors remained secure. As the rescue helicopter returned to the USS *Ranger*, Fischer remained in the water and turned back to continue rescuing the refugees. Fischer's selfless attitude and training put that toddler's life above his own even when he lost essential equipment. Although he now had one remaining fin, he modified his stroke to a single dolphin kick to continue saving lives. The entire crew was later named Aircrew of the Year for the heroic rescue.

Back on the USS *Ranger*, AW2 Neil Packard swapped out with Harcus for the rescue swimmer role. As soon as the SH-3H Sea King was refueled, they headed back to the barge. At this point, there were at least two helicopters alternating hoisting with multiple rescue swimmers in the water saving lives. The aircrew in one of the additional HS-14 aircraft were commanding officer Commander Monte Squires, Lieutenant Junior Grade Walt Beck, Crew Chief AWCS Phill Griffin, and Aviation Rescue Swimmer AWAN Dale Brandt, and in another aircraft Lieutenant George Jacobs, Lieutenant Jeff Bennett, Crew Chief AW2 Kevin McNease, and aviation rescue swimmers AWAN Robert Walker and AWAN Scott Bickerton. I should note that the crew chiefs were also seasoned aviation rescue swimmers but fulfilling the role of hoist operator for this particular rescue.

Back near the barge, Fischer treaded water and encouraged a young female to enter the rough seas. He immediately placed her in a controlled cross-chest carry and began towing her to the extraction area. She was struggling to stay above the water and was panicking due to the horrific trauma unfolding in her world. Fischer decided to give up his SAR-1 flotation to give her a sense of safety. This calmed her down enough to allow the rescue swimmer to continue towing toward the extraction point. Her frantic demeanor then shifted from one extreme to the other, which is when he noticed she started going into shock and losing consciousness. The crew chief hoisted Fischer and the young female up to the hovering Sea King. He quickly disconnected from the rescue hook and went to work on treating her shock symptoms as the pilots flew back to the aircraft carrier.

After the Vietnamese refugees were brought to the USS *Ranger* for medical treatment, they spoke with an interpreter to explain their situation. The barge had been adrift for ten days in heavy seas during the monsoon rains in the South China Sea. The barge had broken loose from where it was moored near a small Vietnamese coastal island. It originally had ten men onboard, but acquired twenty-nine more refugees when their sinking boat came into contact with the adrift barge. The quick reaction and determination of the five aircrews and seven rescue swimmers saved all thirty-nine lives that horrific day. HS-14's commanding officer, Skipper Monte Squires, was quoted in *Rotor Review* magazine: "Open Ocean Search and Rescue is one of the primary missions of HS-14 and the Chargers do it professionally with style. This was an all-hands effort and we're very proud of our squadron's contribution to this spectacular rescue!"

Two years later, HS-14 Chargers were again deployed on the USS *Ranger* to the Persian Gulf in support of Operation Desert Shield / Desert Storm. On August 2, 1990, under the leadership of Saddam Hussein, Iraq invaded the State of Kuwait. Within two days, they had fully occupied the country. In response, U.S. president George H. W. Bush deployed troops to Saudi Arabia and called its allies to join forces to liberate Kuwait. From August 1990 to January 1991, the joint coalition built up their forces as a part of Operation Desert Shield. On January 17, 1991, aerial bombing began against Iraq as a part of Operation Desert Storm. The bombing continued until February 28, 1991, until the U.S.-led liberation of Kuwait.

On February 27, 1991, the day before the war's ceasefire, Lieutenant Jeff Maclay, Lieutenant Steve Swittel, AW2 Jeff Bast, and AW2 Charles Harcus were launched in Lightning 615, an SH-3H Sea King, on a CSAR reconnaissance mission. Four of the six SH-3Hs had been stripped of their anti-submarine warfare sonar gear and equipped with countermeasure chaff/flares, M60 machine guns, Global Positioning System (GPS), downed aviator locator systems, and other tactical components to more effectively operate CSAR missions. Lightning 615 was pursuing four Iraqi commandos fleeing in a 20-foot Boston whaler type vessel near the northeast coast of Kuwait on Bubiyan Island.

A U.S. Navy EP-3 Orion aircraft patrolled the area to provide airborne assistance and communication to the USS *Oldendorf*. The *Oldendorf* was a Spruance-class destroyer, playing a pivotal part in the U.S. response to the Iraq invasion of Kuwait. She earned the Combat Action Ribbon for escorting warships and supporting the naval blockade of Iraq. An HS-12 Speargun 615 SH-3H Sea King attached to the USS *Midway* was also on scene flying aerial armored support during the Iraqi pursuit. Speargun 615 located the small Iraqi boat just offshore of where Lightning 615 was pursing the enemy prisoners of war (EPWs).

Aviation Rescue Swimmers / Door Gunners AW1 Tony Davenport and AW2 Todd Zoldowski lit up the unmanned vessel with 7.62 mm from their helicopter door mounted M60s. After completely disabling the boat, they patrolled the area for any additional aerial support required as the fire-engulfed boat burned and sank.

During this time, Lightning 615 searched inland toward the location where the four Iraqi soldiers fled. They circled the helicopter tightly around an observation tower straddling two Iraqi bunkers. Aviation Rescue Swimmers / Door Gunners Bast and Harcus opened up their M60 on the tower to counter a possible hostile enemy ambush. The 7.62 mm shells rained down from the hovering Sea King as the rounds shredded the sides of the concrete structure. Due to the intensity of military force, weapon dominance, and with no other option but to surrender, the four Iraqi soldiers exited a bunker with their hands up. The pilots immediately landed the aircraft, and the aircrew jumped out to apprehend the EPWs. They searched them for weapons and explosives, then tied their hands behind their backs and placed them face down on the deck of the helicopter. The HS-14 CSAR aircrew then flew the prisoners to the USS *Oldendorf*, where they were interrogated and given clothing, food, and a medical examination. They were then transferred to the USS *Shreveport* for additional interrogation and eventually placed in an EPW camp for repatriation.

In both scenarios, the aircrew were highly trained and prepared to respond to increasingly dangerous scenarios. Aviation Rescue Swimmer School and the real-world training in the fleet ensures the crews are familiar and maintain qualifications for every imaginable situation. They are constantly increasing their emotional intelligence to increase their odds of success, when variables and circumstances change without warning.

* * *

While our Intelligence Quotient (IQ) is pretty set by our twenties, our Emotional Intelligence (EQ) can always be increased. Based on the research of Yale psychologist Peter Salovey, Daniel Goleman (1995) gave five characteristics of Emotional Intelligence:

1. *Knowing our emotions—we need to be self-aware of our feelings and recognize our emotions as they happen.*
2. *Managing our emotions—manage what we are feeling when upset and distressed.*
3. *Motivating ourselves—emotional self-control helps us delay gratification and keeps impulsiveness in check. This is required for self-motivation which can boost performance.*
4. *Recognizing emotions in others—tuning into other's needs and emotions increases empathy, which builds on our emotional self-awareness.*
5. *Handling our relationships—to do so we must manage our feelings as well as the emotions in others, which greatly affects our capacity to lead others.*

Increasing your EQ and self-awareness can help your situational awareness, defined as understanding your environment and knowing what's occurring around you. And further, you may be able to predict future outcomes by applying this knowledge to reduce the risk of injury. Bringing in EQ is also being able to effectively communicate this information to your team where appropriate.

* * *

Situational awareness is a component of emotional intelligence, which is the ability to be acutely aware of our surroundings and respond accordingly. Aviation rescue swimmers hone this skill from day one, as they must always have their head on a swivel, being completely aware of everything within their control. In a trauma scenario, things move and change quickly, and the rescue swimmer must be able to adjust on the fly to positively affect the outcome. In the examples used in this chapter, many hazardous things evolved from the time of initial assessment to the completion of the mission. In all cases, the rescue swimmers and aircrew involved quickly reassessed the needs of those in the water and pivoted to alternative methods to reduce the risk of injuries and maintain focus on saving precious lives.

Helicopter Anti-Submarine Squadron 4 (HS-4) "Old 66" retrieving the Apollo 13 crew in 1970. Courtesy of Michael Longe.

Justin Tate being lowered below an HS-2 SH-60F in the Persian Gulf during Operation Southern Watch. Courtesy of Justin Tate.

HS-14 launched SH-3Hs from the USS *Ranger* to conduct multiple rescues from a sinking Vietnamese barge in the South China Sea. Courtesy of *Rotor Review* magazine.

Brian Dickinson's first search and rescue (SAR) jump from an SH-3 in 1993 during Aviation Rescue Swimmer School in Pensacola, Florida. Courtesy of the author.

Helicopter Attack Squadron (Light) 3 (HA(L)-3) Seawolves' UH-1B Huey gunship coming in to refuel and rearm on a seafloat in Vietnam in 1970. Courtesy of Bill Herbert.

William"Bill" Rutledge posing next to an HA(L)-3 UH-1B Huey during the Vietnam War. Courtesy of William Rutledge.

HS-2's Brian Dickinson lowering supplies from an SH-60F down to the USS *Helena* in the Indian Ocean. Courtesy of Neil Sheinbaum.

Brian Dickinson saying goodbye to JoAnna before flying to the USS *Constellation* for a six-month deployment. Courtesy of the author.

Brian Dickinson firing 7.62mm rounds from the M60 mounted in the HH-60H during a training flight in the Persian Gulf in 1997. Courtesy of the author.

James Buriak with wife, Megan, and son, Caulder, reunited after returning from a six-month deployment. Courtesy of Megan Buriak.

HS-7 lowers a rescue swimmer from an SH-3H to the water to save the crew of the USS *Bonefish* after it caught fire off the coast of Florida. Courtesy of the Department of the Navy.

Aviation Rescue Swimmer School graduating class 9319 conducting SAR jumps in Pensacola, Florida, in 1993. Courtesy of the author.

An SH-60F from HS-2 crashed during a humanitarian assistance / disaster relief (HA/DR) support mission in Indonesia after the 9.3 magnitude earthquake in Sumatra in 2004. Courtesy of Cory Merritt.

An aviation rescue swimmer jumps from an MH-60S at 10 feet and 10 knots during a training exercise. Courtesy of Cale Foy.

HSC-22 MH-60S aviation rescue swimmers training with U.S. Navy special warfare combat crewmen (SWCC) assigned to Special Boat Team 20 in Norfolk, Virginia. Courtesy of Jaxson Ingraham.

Brian Dickinson with other exhausted and hungry Survival, Evasion, Resistance, and Escape (SERE) School students after returning from a week in the field during the fall of 1993. Courtesy of the author.

Whitney Warren gearing up as a SERE instructor during a final evasion run in Warner Springs, California. Courtesy of Whitney Warren.

Shawn Porter and Helicopter Combat Support Special Squadron 5 (HCS-5) preparing to launch inside Iraq for strike rescue and Naval Special Warfare (NSW) support. Courtesy of Shawn Porter.

Naval Air Station (NAS) Whidbey SAR UH-3H makes an approach over a river in Washington State. Courtesy of Marty Crews.

Joe Sutherland conducting cliff rescue procedures while attending the Coast Guard Advanced SAR School in Astoria, Oregon, in 2000. Courtesy of Joe Sutherland.

Cory Merritt helping evacuate an injured victim during the HA/DR support mission in Indonesia in 2004. Courtesy of Cory Merritt.

An MH-60S from HCS-5 launching for a combat search and rescue operation in Iraq. Courtesy of *Rotor Review* magazine.

Aviation Rescue Swimmer School candidate performs a body sweep to check for entanglements or injury in the pool at NAS Pensacola. Courtesy of Wikimedia.org

An MH-60S from HCS-5 fast-roping special operations personnel on a rooftop in Iraq during an insertion and NSW support operation. Courtesy of Shawn Porter.

Tim Hawkins hoisting with a victim to safety from the 2005 Hurricane Katrina disaster. Courtesy of Tim Hawkins.

Naval Air Station Whidbey Island (NASWI) SAR conducting medevac training in the Cascade Range. Courtesy of Drew Worth.

The 9D6 helo dunker, which replaced the 9D5. Courtesy of Jay Shropshire.

The 9D6 helo dunker submerged and inverted during a daytime training evolution for egressing a crashed helicopter. Courtesy of Jay Shropshire.

The 9D6 helo dunker submerged and inverted during a nighttime training evolution for egressing a crashed helicopter. Courtesy of Jay Shropshire.

Melissa Dixon from NAS Whidbey SAR operating on the summit of Mount Baker in Washington State. Courtesy of Melissa Dixon.

Melissa Dixon rappelling from an NAS Whidbey SAR MH-60S on Mount Baker in Washington State. Courtesy of Melissa Dixon.

Chapter 6

Attention to Detail

On a hot summer day in the late 1990s, I was on military leave visiting my family up in beautiful Rogue River, Oregon. It was one of those relaxing days that brings you back in time, where you don't have a worry in the world. My older brother, Rob (Navy veteran) and I packed our fishing poles and gear, then headed to the river to catch some dinner. We drove north for 7 miles through Grants Pass and found a gravel road, parallel to the raging Rogue River. We followed the windy road, enjoying the view and catching up with our typical brotherly nonsense talk for a few miles as we distanced ourselves away from civilization. I stopped my 1995 Toyota 4x4 to manually lock the hubs of the 33 x 12.5 tires. From there on, we were off-roading the final section over the unstable, rocky shore. We found a decent flat area to park on the east side of the river, with a calm section to cast into.

The whitewater rapids on the far side looked to be class III–IV, stretching for a few hundred yards. The steep gradient of the riverbed causes an increase in water velocity and turbulence, which leads to unpredictable and violent rapids. It can be a mystifying scene to watch, like spending hours staring at a majestic waterfall. You tend to get lost in your thoughts, and all the world's problems temporarily vanish. With the green mountains and trees, the ferocious river made for a perfect backdrop to enjoy an afternoon of fishing. It's those types of days that you don't even care if you get a bite or not; as the saying goes, a bad day of fishing is better than a good day at work. And that was the case for us as we cast out and patiently waited for the fish to gain interest in our multi-color power bait.

An hour of peaceful fishing flashed by in what seemed like a few minutes. As I was zoning on the passing water and enjoying the sweet smell of fresh water mixed with wild blackberries growing near the shore, an unusual movement caught my peripheral vision. Several hundred yards upriver near a bend, a woman was running along the uneven shoreline waving her arms and frantically screaming. That's when I noticed three heads bobbing through the massive rapids about 60 feet across on the far side of the river. A surge of adrenaline flooded my body as it automatically went into rescue swimmer mode. Without hesitation, I dropped my fishing rod and tore off my shirt as I sprinted up the riverbank. Quickly calculating the environmental conditions, the victim's distance, speed, and my ability to catch them, I dove into the frigid 64-degree water. The shock of the water temps matched with my heightened adrenaline charged my system into a channeled energy with a single focus, saving lives. I used the American crawl technique at my full strength, taking a small breath every few strokes and ducking into the crashing rapids of the epic ebbs and flows of the river.

The fast current quickly swept me downriver, but I didn't fight against it. I had estimated a point of entry higher up the bank, knowing the current would propel me down as I conducted my swift water rescue approach. I was able to use the rapids to provide momentum by entering high on the crest and riding down to the trough as I crossed toward the survivors. By the time I reached the far side of the river, the three had spread out from each other. I assessed the situation as I got closer and noticed a very young girl struggling to keep her head above the water. I immediately prioritized her and established verbal contact. I spoke in a calming but commanding voice to gain her trust while not causing additional panic. She was a lot smaller than the civilian and military adults I'd trained or rescued in the past. Continually communicating, I physically turned her around so she was floating on her back with me underneath, supporting her with my left side. I brought her into a cross-chest carry and attacked the rapids, doing all I could to keep her above water and out of harm's way. I could hear her crying and coughing through the loud sound of the river but didn't slow down since I needed to get her to safety before we were both swept further downriver. I hit several large but smooth submerged boulders as I fought the waves, reversing my method from earlier of riding up and down the rapids on our return. When we both reached the dry east bank of the Rogue, I quickly assessed her condition and then told her to sit, breathe, and wait there while I went back for the others.

The other two were still caught in the unrelenting river. They were trying but failing to swim toward shore which was causing overwhelming exhaustion. I quickly dove back in and swam toward the two, who were still spread apart from each other. I could tell they were fatigued and panicking. I reached a middle-aged-looking woman first and let her know I would take her next. I also yelled toward the man for him to try to remain calm as I would be back for him. The middle-aged woman was submissive as I pulled her back across the river. We reached the shore farther down from the little girl, but the exhausted woman ran as soon as she reached dry land to her awaiting daughter. My brother had just finished reeling in our fishing lines, stowing our poles, and was making his way to the casualty collection point as I reentered the water for the final victim. A casualty collection point is a safe location used for assembly, triage, medical stabilization, and if necessary medical evacuation. In our case it was an area to collect the victims to check their vitals and treat any injuries.

I returned to rescue the remaining free floater, an older man. He had found a calm eddy on the far side of the river where he was treading water. An eddy is where the water flows back upstream after a deflective obstacle, like a submerged boulder. When I reached him, I could see he was completely exhausted with no more fuel in the tank. I grabbed hold and kept him afloat and then spoke calmly about how he could relax now as I would tow him back. I carried him in a cross-chest carry and side-kicked, using my right arm to pull at the water with long arched strokes. The man was bigger than the other two put together and I was fatigued, but I knew I had to keep going as I could rest when the rescue was completed. My knees, shins, and knuckles ached and bled from striking the large, submerged rocks. Once we reached hip-deep water, I stood us both up and slowly guided him to shore. I dragged him up on the sandy bank 20 yards down from the casualty collection point, where he collapsed, staring lifeless into the sunny blue sky. His wife, in shock, ran down from the riverbank and sat down next to him, holding his hand while uncontrollably sobbing. Peripherally I noticed the mom hurrying her daughter away from the scene in shame without a word or even looking back. The traumatized little girl looked uncomfortable and confused but obeyed her mom as they disappeared around a bend in the river. I turned my complete attention to the man lying in the dry sand. Checking his vitals, his breathing and heart rate was off the charts. He just needed to lay there for a while and catch his breath and calm down. I remained with him for a few minutes as his body regained some normalcy and he was able to consciously interact with me and his wife.

After he relaxed a bit, he explained what had happened. Apparently, the little girl was playing too close to the river's edge when she was swept away by a rogue (no pun intended) wave followed by deadly undertow. An undertow is the undercurrent moving offshore when waves or rapids are moving toward shore. Undertow can be very lethal, a major contributor to the four thousand drowning deaths in the United States each year. Additionally, the little girl didn't know how to swim. The mom heard her panicked scream and did what any parent would do: She jumped in to save her even though she wasn't a strong swimmer. The older man I was attending to was a retired fireman enjoying a peaceful afternoon with his wife. Relaxing in a frayed vintage web folding lawn chair, reading a World War II novel, he witnessed the traumatic event. And like the hero he was, he disregarded his own safety and jumped in the raging river to help. Unfortunately, the river overpowered them all, and what started as one victim quickly grew to three.

He and his wife were extremely grateful and blown away that I was a U.S. Navy aviation rescue swimmer, who just happened to be in the right place at the right time. He was extremely exhausted and emotional as he sat in the dry sand while covering his face with his shivering pruned hands. He felt it was God watching over them all. I never saw or heard from the woman or child, but that's common when people get into life-threatening situations in which embarrassment overshadows gratitude. I've rescued a few victims for whom was the case, and I've rescued a few whowere overly grateful. I've heard the same responses from other rescue swimmers and first responders. Ultimately, it's not about receiving praise; it's about unconditionally saving lives. And you never know when you'll have the opportunity to run toward the fight. Rescue opportunities rarely happen when you expect them, but you always have the opportunity to respond to them. That's the difference when you have a mindset of serving others and saving lives.

AWRCM Michael Ousley served over thirty-one years as an aviation rescue swimmer. His response to my question around the need for such intense training was: "The real reason is 'simplicity and primacy,' repetition in 'no win' scenario training leads to one's ability to process vast amounts of stimulating information and make a choice to take action. Fundamental 'steps' and repetition in training creates opportunity during high stress situations in order to 'slow down' and acknowledge contingencies to remain on task in order to complete the next indicated step, no matter what that might be. Rescue Swimmer development has thousands of examples of impossible scenarios to pull from and candidates are exposed to ever increasing levels of 'degraded operations,' including screaming, bleeding, unresponsive, debris, and petroleum filled pockets of hazmat to navigate. All to be negotiated with a singular purpose of 'rescuing' the survivor/patient. Each event followed with critical debrief highlighting a physical capability with critical thinking layered with situational awareness. The exact recipe for building 'predictability' of an individual operating independently of a team. A team can modify actions to adapt to what an individual is reacting to while the individual can reliably 'expect' a response for support from the team."

Another more recent example of being in the right place at the right time and responding to the task occurred on May 2, 2021, off the coast of Point Loma, California. On that picturesque Sunday morning, AWS1 Cale Foy was out hiking with his family near the Cabrillo National Monument. He noticed out in the high surf an empty-looking vessel going through a seaweed patch. Something didn't quite look right, so he and his family made their way north on the beach line, which led closer to the water. The boat slammed into the offshore rocks and disintegrated, tossing people, luggage, and debris into the water. Without hesitation, Foy took action.

He and another service member standing by assessed the situation. There were 5-to-8-foot waves crashing into the rocks and coral. They were both in their hiking clothes, consisting of pants, T-shirt, and boots. Rescue swimmers typically enter the water with fins, mask, snorkel, helmet, wetsuit, harness, and emergency flotation. However, once Foy saw the victims flailing in the frigid 60-degree ocean, he immediately entered the water and shifted his mindset to putting the victim's needs above his own to try to save as many lives as possible.

The sinking vessel was a migrant smuggling boat, with over thirty Mexican migrants onboard. Foy and the other service member assisted a group of victims to shore and then headed out beyond the surf to help the others. The high seas made it difficult, but Foy found a rip current that helped pull them both through a safer passage through the crashing waves. The water surface had a thick layer of fuel, fiberglass, wreckage, and luggage. Visibility was limited due to the white foam of the crashing waves and excess fuel in the water.

Foy noticed the top portion of the cabin was decapitated from the boat but floating on its own. He decided to use this as a makeshift mass casualty collection point. He zigzagged through the wreckage, pulling the wounded and hyperthermic victims to the floating cabin roof. Some were calmer than others, but all dealing with shock. One man panicked, forcing Foy to utilize his life-saving training techniques to gain control and calm the man down in order to tow him to safety. There were several people thrashing in the water, but there were also a few that were face down that had unfortunately succumbed to the conditions.

As Foy was pulling victims to safety, lifeguard and Coast Guard watercraft and rescue personnel arrived to take over. Foy helped hoist two people into a rescue boat, then climbed in himself and began performing CPR on one of the victims. He continued his efforts through the choppy ride to a dock in the Port of San Diego. The dock became the new casualty collection point, where all the victims were gathered for medical assessment. Foy was happy to see all the people he had towed to the floating cabin had been relocated there. He continued performing first aid on the victims, until he was relieved by other medical personnel. Three people died, but twenty-nine were alive. Foy had been in the cold water for ninety minutes, but he doesn't recall the water being cold. It wasn't until after the traumatic event that he began shaking from the low temps and emotional letdown.

"If someone else's life was in play, I'd be willing to endanger mine, absolutely," Foy said. "When opportunity arises, there's no rescue swimmer who won't jump to the cause."

AWS1 Foy was awarded the Navy and Marine Corps Medal for his willingness and swift action to save lives. He was also named the U.S. Navy Sailor of the Year for his actions. The traumatic scene he swam toward to help in a time of dire need weighed heavily on him. It's something that many in the field don't want to admit or be labeled, but everyone is dealing with it. His courage to acknowledge his feelings, talk about and process it, and lean on his rescue swimmer community has helped him find peace.

* * *

Trauma is the Greek word for wound, and trauma can be a physical wound or an emotional wound. The psychological reaction to an extremely stressful event such as war, natural disaster, or abuse is considered post-traumatic stress disorder (PTSD). Symptoms can include recurring nightmares, anxiety, flashbacks, difficulty focusing, depression, anger, agitation, numbness, or suicidal thoughts. If you are struggling with any of the symptoms of PTSD, please seek help! If you are suicidal, please call the suicidal hotline: 988!

* * *

"On duty, and off duty in this case, AWS1 Foy definitely embodies the motto and never second guessed himself about what he had to do," AWC John Conant wrote, "placing himself before others as he continuously does at the command for our Sailors."

"Right place, right time, right Naval Aircrewman," said Foy's commanding officer at the time, Capt. William Eastham, in a 2021 statement. "This was a one-of-a-kind off-duty rescue, and thinking about the conditions he encountered without any of his prototypical SAR gear—in just a pair of jeans, a T-shirt and hiking boots—it really ups the ante. Courageous and immediate action to be sure. The Naval Aircrewman motto, 'So others may live,' never felt as true as it does today and AWS1 Foy's actions certainly embodied just that."

Worth mentioning, a year prior to the above rescue, AWS1 Foy was also a part of a Navy first. In July 2020, the USS *Bonhomme Richard*, a multipurpose amphibious assault ship, caught fire while moored at Naval Base San Diego. The fire grew too hot and widespread for crews to battle the blaze aboard the warship. Helicopter Sea Combat Squadron 3 (HSC-3) was called to drop buckets of water to cool the superstructure and flight deck. A helicopter bucket is a specialized bucket suspended from a cable below the helicopter. The crews submerged the buckets in the San Diego Bay and released them over the burning ship. The squadron flew continual operations, dropping fifteen hundred buckets of water. Foy and his pilots and aircrew were able to reduce the shipboard fires enough to allow ground crews to take over and eventually extinguish the fire. Unfortunately, the extensive fire damage resulted in a total loss of the ship. It's just another example of the diversity of the skills a Navy aircrewman must have.

* * *

After a full day of training, there isn't much time to relax during the evenings. Following dinner at the NAS Pensacola base galley, the students go back to their barracks and starch and iron their uniforms. They then polish their boots until they look like glass. The candidates need to get to sleep and wake up early to stand at attention for inspection outside the Aviation Rescue Swimmer Training Facility. From the moment they step foot at the training facility until they leave, the instructors hammer them on their attention to detail. From a flawless uniform to meticulous training techniques, to exercise cadence, everything is closely evaluated and punishable. And no matter how ridiculous it seems, each lesson is taught for a reason. This helps the students become efficient and effective aviation rescue swimmers in the fleet.

* * *

Learning and maintaining attention to detail may be easier or difficult based on one's personality. Since certain jobs require this skill, it can be helpful to take a personality test to better understand your unique strengths and weaknesses. The more you understand yourself and how you're wired, the better you can understand your capabilities. This also helps you better understand others and have grace for them as well as yourself, as we are all created uniquely.

People typically either have a strong attention to detail or they see the big picture, which can impact their focus on the smaller components. Both types are necessary for a team and help complete a unit when the strengths and weaknesses of individuals complement one another. You can also work on increasing your attention to detail by eliminating distractions, not multi-tasking, and engaging in activities that practice focus like active listening skills, self-awareness, and situational awareness. Some people also find they can improve their focus by doing self-care practices such as exercise, deep breathing, mindfulness, meditation, and organization. These all reduce stress as they help improve your working memory and attention to detail.

* * *

Besides being strong, confident swimmers and experts in water rescue, the rescue swimmer candidates learn advanced first aid to be able to provide emergency medical support on land and in the helicopter. Later in the fleet, some aviation rescue swimmers also have the opportunity to further their medical knowledge by attending Emergency Medical Technician (EMT) training. A lot of the rescues in the fleet are medevacs, where the aircrew fly to retrieve injured personnel, critically hurt victims, or body recoveries. This includes any scenario you can imagine, from landing on the back of a frigate to extract a heart attack victim, flying an injured sailor from an aircraft carrier in the Persian Gulf to Camp Doha in Kuwait, to hoisting a Navy SEAL suffering from diving injuries from a submarine to be transported to medical facilities.

Each day the students spend countless hours in the classroom in-between land-based and water-based physical training and procedures. They are taught and evaluated on the fundamentals of first aid and CPR, then moved to advanced techniques and treatments. The candidates must know the symptoms, procedures, and equipment forward and backward. Each aspect of the training is carefully evaluated under high-pressure scenarios. Any failure means you are disqualified from completing the program. This makes sense since it does no good to save a life if you can't keep them alive!

U.S. Navy Corpsman HM1 Kevin Frank conducted this portion of my training for class 9319. He was highly talented and took the training to the next level with his skilled utilization of moulage. Moulage is an artistic technique where you utilize gory cosmetics and supplies to create mock injuries on real humans for emergency medical response training. The artificial injuries are legit! You can essentially turn a victim or victims, in our case, into horrific trauma scenes. We saw open fractures with bones exposed, sucking chest wounds from bullet wounds, and arterial bleeding with blood squirting everywhere. It's as real as it gets without being real!

On March 9, 2002, HM1 Frank was aboard Angel One, a CH-46, with Marines assisting the U.S. Coast Guard in a search for survivors of a downed civilian helicopter 30 miles southeast of Savannah, Georgia. The civilian helicopter was carrying two workers on a Marine Corps project, Tim Potts and Keith Laney, who both perished in the crash. During the open ocean search, one of the crew noticed something in the water that looked to be one of the victims. When Angel One turned back toward the possible survivor it experienced a major malfunction, causing it to crash into the waves below. After the two main rotors violently ripped off from striking the ocean, the fuselage filled with water, turned over, and sank. After the commotion stopped, the five-person crew egressed from the sinking helicopter. Word is that as Frank made his way to safety, he noticed the crew chief was hung up by the long cord for internal communications. Frank broke the number one rule of helicopter egress and went back in to free the crew chief from the fouled cord. The crew chief eventually surfaced with the remaining aircrew minus HM1 Frank. His body was recovered days later by U.S. Navy divers. HM1 Kevin Frank died a hero, doing exactly what he loved doing, saving lives.

The rescue swimmer students train in the classroom to become proficient and desensitized to working with the mangled victims. They learn to not show emotion as their reaction can cause a victim to panic and go further into shock. They learn about all the medical gear and procedures under close evaluation. Once they have the controlled classroom training complete, the instructors introduce moulage to the multi rescue scenarios. For overwater rescues, aid begins in the helicopter since it's too difficult to administer aid in the water. Depending on the situation they may give rescue breathing or some basic support in the water or as they're hoisted up, but for the most part there's not a lot they can do until they're in the aircraft. (The search and rescue model manager [SARMM] recently removed in-water rescue breaths from the standard operating procedures to expedited extractions during airborne rescues.)

After a packed day of land, water, and classroom training, the rescue swimmer candidates will finish the afternoon with multi rescue scenarios. Everyone either sits in full gear on the side of the pool observing and waiting for their own turn or they conduct other specific training or conditioning with other instructors. As described in chapter five, the multi scenarios include extreme situations meant to stretch the students to their maximum ability. They will leave the training with a fresh concept of their limits, which are pushed far beyond what they previously imagined.

During my final multi, I was hoisting up with my last survivor, an F-14 Tomcat pilot. Once I reached the door and we were both lowered into the cabin, I quickly assessed the life-threatening chaos that awaited my next evaluation. It looked like a war zone, which was the intention, with two bloody victims moaning on the ground with another frantically pacing. Victim One was frothing at the mouth and choking on his own teeth. The creative advanced first-aid instructor used white Tic Tac candy to simulate busted out teeth. Victim Two was moaning from a bullet puncture injury: a sucking chest wound as well as blood squirting across the cabin from an arterial laceration in his thigh. I would have prioritized the most life-threatening injuries first, but instead I had to deal with Victim Three, who was clearly suffering from shock.

The shock victim had no visible injuries, but he was distractingly poking the other victims, saying he was trying to help. When confronted he decided to attempt taking his own life by trying to open the cabin door to jump out of the moving helicopter. I attempted to use my available resources by asking if the crew chief could help, but he said he was busy. That would have been too easy. However, if I didn't ask, I would have lost points for neglecting to tap into my available resources. Knowing I was on my own, I quickly closed the cabin door to prevent anyone from jumping or falling out and then detained the shock victim. I strapped him down to prevent him from harming himself or anyone else, then I turned my attention to the dying victims.

I started working on the arterial bleeder and then the other victim went into cardiac arrest. I placed my knee on the bleeder's artery to stop the spurting blood and reached over to sweep the other guy's mouth to clear it from teeth and obstructions. I then gave a couple breaths while checking his pulse and then began chest compressions. In the deafening helicopter noise with extreme vibrations, it's imperative to be certain of a positive or negative pulse. In extremely cold weather, it can also take up to a minute to feel a pulse. Every second is crucial and if done incorrectly can have devastating effects, since chest compressions on a beating heart can cause a heart palpitation. I continued chest compressions for a minute before my crew chief became "unbusy" and took over on CPR so I could help the bleeder.

I applied a tourniquet and wrote a T with the time on his forehead from his own blood. This is to inform the next medical technician he had a tourniquet and the time it was set, which helps them determine the likelihood of saving the limb. I kept checking in with the other victims, including my strapped-down chatterbox in the corner who was determined to distract me. After the arterial wound was treated, I worked on his sucking chest wound. I created a three-sided seal with a plastic wrapper from the medical kit to prevent a collapsing lung. The seal is meant to prevent air entering in from the new pathway, but still allows extra air to go out. After a while this can get saturated or plugged, requiring purging the wound with a clean finger.

When Victim Two was deemed stable, I relieved the crew chief until Victim Three had a pulse and was breathing on his own. I conducted a head-to-toe check and found a battle sign, which is bruise behind the ear, indicating serious head trauma. I placed a neck brace on him and ensured he was strapped securely in a litter to prevent movement. Throughout all of it I was communicating what I was doing and their status to the pilots, who relayed the information to a nearby hospital. All patients were stable, and the instructor concluded this portion of the multi exam.

This was one of many multi rescues coupled with advanced first aid. I didn't do everything perfect, but I would learn and grow from my minor mistakes. With so many moving parts, emotions, noise, and distractions it is critical to continuously maintain the highest standard of operation. The instructors watch carefully to see how the students respond to difficult scenarios, as it's not always the solution that matters but how they get there. It shows their acute attention to detail in a completely chaotic situation in a relatively controlled training environment. If they could handle it in training, then there's a good chance they'll be able to handle it in the fleet. No instructor would rightfully sign off a candidate if they weren't consistently exceeding the program's high standards.

Aviation Rescue Swimmer Drew Worth has conducted over one hundred rescues in his nearly three decades of military service. Stationed in the Pacific Northwest at Naval Air Station Whidbey Island (NASWI), he was regularly launched on mountain rescues in the unforgiving Cascade Range. The Cascades are a major mountain range in western North America stretching over 500 miles from British Columbia down to Northern California. They include an active volcano chain of mountains as well as non-volcanic peaks. They are the pinnacle destination for outdoor extremists as well as novice adventurers. Each season presents intense natural elements to challenge the highest skilled individuals. However, with the extreme challenges comes risk, with thousands of accidents ranging from minor to critical occurring each year.

NASWI, located in Oak Harbor, Washington, was created in 1941 almost a year prior to the United States entering World War II. The Office of the Chief of Naval Operations (CNO) requested the commandant of the 13th Naval District to find a location for refueling and rearming Navy patrol planes operating in defense of the Puget Sound. Oak Harbor was originally rejected as the location for the base due to the high mountains, inaccessibility, and bluff shorefront. However, it was quickly reconsidered due to a long strip of land that would be converted into an airfield. The first stationed aircraft in NASWI were the carrier-based fighter Grumman F4F / F6F Wildcats, Lockheed PV-1 Venturas, Douglas SBD Dauntless dive-bombers, and Consolidated PBY Catalina and Martin PBM Mariner seaplanes. The base has evolved through the years as a result of world conflicts and military needs. Today there are twenty active duty and three ready-reserve U.S. Navy squadrons based at NASWI. Their missions are electronic attack squadrons, maritime patrol, fleet air reconnaissance, fleet logistics support, and SAR.

The Naval Air Station Whidbey Island Search and Rescue (NASWI SAR) team was created in 1960, using two Sikorsky HRS-2 helicopters. This was a two-year trial period to see if there was a permanent need for rotary support. Apparently, there was a need, and it has grown significantly in complexity and capabilities, as NASWI SAR is considered one of the elite rescue teams in the U.S. Navy. They currently fly Sikorsky MH-60S Seahawk helicopters to perform twenty-four-hour maritime, inland, and mountainous rescue support for military and civilian personnel.

During an unusually warm winter in 2009, Navy pilots Lieutenant Jeremy Ethridge and Lieutenant Casey Bruce, AWS1 Drew Worth, Crew Chief AWS1 Eric Deburkarte, and Search and Rescue Medical Technician (SMT) HMC Gregory Highfill, were dispatched from NASWI to search for a climber in distress. Due to the fluctuating warmer temps, the Cascade Range saw more recreational activity than normal for the winter months. The MH-60S Seahawk helicopter departed the Puget Sound island and flew south near Mount Rainier to investigate a fallen ice climber. Mount Rainier is the highest glaciated volcano in the Lower 48 states. Towering at 14,411 feet, it is a popular mountain for climbers, skiers, outdoor enthusiasts, and day-trippers. People travel from all over the world to either make an attempt on its cratered summit or even just to take a selfie from the Paradise or Sunrise Visitor Center parking lots. I have personally climbed Mount Rainier and the surrounding peaks multiple times to train for higher mountains like Denali and Everest. Although there are plenty of more interesting mountains in the North Cascades, in my opinion, it's the pinnacle peak that most adventurers add to their bucket list. Each year I try to help fulfill a few dreams with expeditions on Tahoma, as it was named by the Northwest Native Americans, leading groups including former military, Navy SEALs, and NFL players. However, with all its popularity comes risk of injury and natural disaster. Each season brings multiple SAR missions to the mountain from the park rangers, guides, and local Navy and Army Reserve Search and Rescue. To date, Rainier has claimed the lives of over four hundred individuals.

NAS Whidbey SAR received the message of the fallen ice climber, relayed by the victim's climbing partner who was able to descend the mountain to get cellphone reception and get an emergency call out. The emergency dispatchers alerted a ground rescue crew as well as called Whidbey SAR. When the MH-60S arrived on scene, they performed a search pattern through low visibility conditions in the fog and clouds. They mainly relied on their instruments to ensure they remained a safe distance from the mountain as the aircrew searched through the cabin windows to avoid any potential rotor strikes from trees or from the snowy mountain itself. The ground rescue crew was able to hear the helicopter and vector them in closer to their location, which was a few miles farther south of Rainier. The helicopter SAR crew made visual contact with the fallen victim due to his bright-colored jacket, which stood out from the surrounding snowy peaks. The pilots checked the winds and adjusted their position and power for an approach as the aircrew in the back rigged for a rappel rescue. The MH-60S is equipped for rappelling or hoisting, but the fastest method to deploy at 300 feet and mitigate risks of avalanches is to rappel. The crew chief, Deburkarte, created a self-equalizing anchor system, attaching webbing to locking carabiners, which attached to support rings and pulleys in various locations in the cabin of the helicopter. The rappelling rope was then attached via a figure-8 knot to two counter-facing locking carabiners. Deburkarte tossed out 300 feet of rope as Worth and Highfill prepared their harnesses and rescue equipment for insertion.

The SAR helicopter pilots were extremely careful of their hover approach and aircrew insertion due to the high avalanche danger. Avalanches are caused when something triggers instability in steep slopes typically 30 degrees or steeper, with a weak layer of snow covered by a heavier layer. The tremendous sound and force of the helicopter's rotors could easily trigger a massive avalanche, turning the rescue into a recovery. The MH-60S kept a high hover as Worth and Highfill connected to the rope one at a time to rappel down to the injured victim. The ice climber had found a perfect vertical ice cliff to challenge his abilities. Unfortunately, it was so steep that Highfill needed to rappel to a higher elevation rather than directly to the victim. The rope heated up as it rapidly fed through the medic's steel figure-8 friction device attached to his harness. Highfill carried a first-aid kit and touched down on a snow embankment about 200 feet above the victim. Leaning over the edge, while holding a tree for support, he yelled down to make verbal and visual contact with the survivor. From that distance it was impossible to assess the climber's injuries, but he radioed to the above helicopter for Worth to bring additional equipment.

Worth attached his rappelling device to the rope and, carrying the rescue litter, snow anchor, and rappel equipment, stepped out from the cabin door, and slid down the rope to meet Highfill. They both investigated their surroundings to see what available natural resources they had to solve the puzzle of getting down to the helpless victim. There was a lone, large-trunked conifer close to where the rescue crewmembers stood, which made for a perfect anchor point. They double-wrapped webbing and counter-faced locking carabiners to hold the rappelling rope. After testing its safety and durability, one by one they rappelled down the snowy cliff face. Both rescuers wore crampons—metal-spiked traction devices attached to footwear to improve safety and stability for ice and snow travel—to maintain purchase on the snow and ice but had to be diligent to not pierce the rope. They both remained attached to the rope for safety once they reached a small snowy platform near where the climber had landed. They remained calm and unfazed by what they found and began communicating with the surprisingly calm victim.

Lieutenant Ethridge and Bruce decided to relocate the helicopter to a safer location a thousand feet below from the rescue. They found a flat clearing and were able to land to preserve fuel and reduce sound. They were far enough away that the rotor wash and audible thumping of the main rotors hopefully wouldn't trigger an avalanche up higher, which would have certainly wiped out the victim and rescue crew. With the rotors spinning, they maintained radio contact with Highfill and Worth, relaying their status to emergency dispatchers and ground rescue crews.

The injured, shocked, and freezing ice climber explained to the two Navy rescuers what had occurred. He was placing protection for his final crux to the summit, which was capped by a massive cornice, an overhanging mass of hardened snow and ice at the edge of a mountain precipice, typically created from high winds, heavy snow, and freezing temperatures. When the climber struck his ice axe into the hanging layer of ice, a section the size of a car broke loose, and he fell over 800 feet. He bounced off ice, rock, and trees as he accelerated down the mountain. He felt completely out of control and the panic and fear nearly caused him to black out before he came to a sudden stop. He recalls a whiplashing halt and all the air being knocked out of his lifeless body when he slammed into the top of a Douglas fir, which completely skewered him through his back and exited through his side. The tree must have hit bone and redirected inside his body since it was bent inside of him. He laid pinned, staring at the sky for hours as he waited to either die or for help to arrive. Highfill and Worth needed to cut the climber's clothes and equipment off to assess the true damage. Worth recalls the in-shock victim worrying more about his clothes and gear than his actual injuries. He calmly reminded the survivor that his gear could be replaced, but they needed to stabilize him to be able to hoist him out.

As they carefully cut and removed articles of clothing, they did their best to keep their composure as they planned out the next steps. A first responder's facial expression when seeing the extent of injuries can send the victim into shock. Since the victim themselves can't normally see or process what has happened, their first indication is how they see others react, which can then trigger their own heightened response. Rescue swimmers and medics are trained to maintain a flat affect facial response to not make matters worse. Highfill and Worth saw the 4-inch tree trunk, stripped of its upper branches and needles, stuck through the victim's body. The sweet smell of freshly cut pine trees countered the terrifying reality of what they were viewing. The impaled tree was not only keeping the victim from bleeding out, but it also prevented him from falling another thousand feet to his certain death. The victim was extremely cold from his long exposure to the elements, which was probably a good thing, as it slowed his heart and reduced the hemorrhaging. This lone Douglas fir was a gift from heaven, saving his life from a couple different perspectives. And the SAR team was going to cut him a souvenir, which would ultimately save his life.

Highfill provided a low dose of morphine to the climber as Worth cryptically communicated their plan to the MH-60S crew, since he knew the victim would hear his radio transmission. Highfill and Worth then diligently carved out a more stable snow shelf from the side of the cliff with their snow shovels. They stomped it down with their boots to test its firmness to hold the three of them plus equipment. It was a necessary step to create a safe platform to stabilize the victim in the rescue litter, which Worth assembled and anchored to himself while still connected to the rope and tree above. While he worked, the helicopter crew began preparing for a rescue litter medical hoisting evacuation. They lifted up into a hover and flew a nearby pattern, waiting for the rescue crewmen to call them in for extraction.

The next procedure was the only real option for saving the victim's life, and both crewmen agreed on what needed to happen. Worth pulled a small handsaw from his mountain SAR equipment bag and handed it to Highfill. They would need to cut the tree on both sides of his body, leaving the wood trunk impaled until they reached a medical facility. The number one medical rule of impaled objects is to leave them. Removing them damages nerves and blood vessels. And in an extreme case like this, removing the tree completely would possibly remove organs and the victim would bleed out and die within minutes. The SAR medic sawed through the top portion of the tree piercing through the victim's side. With each pull, the ice climber moaned with agony as the evergreen conifer tore back and forth through his body. The crewmen did all they could to keep him stable, but the slightest movement of the bayonetted object radiated like fire through his injured torso. Highfill packed and bandaged the remaining 2 inches of tree sticking through him. Worth held the climber steady as Highfill started slowly sawing through the lower portion of the tree. He kept about 4 inches of the trunk on the backside of the victim, which he would bandage in place once cut free.

This would allow for easier access and delivery once they got him to a medical facility. He found his rhythm and sawed efficiently minimizing vertical movement. What took a few minutes felt like a few hours to the victim, but Highfill was able to cleanly cut the climber from the tree. Worth recalls how little bloodstained the snowy ground was, both because the tree had essentially caused and sealed the wound and the freezing temperatures, which helped slow the hemorrhaging process.

Both gently lifted the ice climber and moved him laterally to the rescue litter. They then performed first aid, dressing, and securing the lodged element with battle dressings. They removed the climber's harness from his waist and legs, using it as a buffer around the tree stump sticking out his back. Worth securely strapped him into the litter with all the proper restraints and then radioed the MH-60S that they were ready for extraction. The pilots pulled into a low-and-steady hover as Deburkarte lowered the rescue hoist. They would need to work diligently, as they were still at high risk of avalanche danger, especially being later in the day as the snow softened from the afternoon sun. Highfill picked up the grounded rescue hook and connected the litter sling cables to the large hook, connected his lifting V-ring to the hook, and gave the crew chief a thumbs up. Deburkarte raised the SAR medic and litter up as Highfill tried using his arms to prevent spinning. Due to the vortex winds from the rotor wash and gulley they were hoisting, it was impossible to not spin. Debris from the ice, snow, and trees ripped through the air around them as they wound their way up to the MH-60S. The crew chief got his first view of the impaled victim as he hoisted them into the helicopter cabin. He knew his team faced a unique challenge but did a great job in keeping him alive and safe for extraction.

While the other two were being hoisted up, Worth climbed back up the 200-foot vertical cliff to retrieve his rappelling anchor gear and rope. During that time, Deburkarte got everyone onboard, and the pilots flew a lap around the area to ensure they didn't kick off an avalanche on Worth. When he had his gear packed, he radioed the circling helicopter, which returned and hoisted him up. The injured climber was unhappy on the uncomfortable flight to Harbor View Hospital in downtown Seattle, due to the high winds and bumps, but Highfill was able to administer more medication once they warmed him up. The injured ice climber would survive with a gnarly scar due to the quick actions of the NAS Whidbey SAR team. The entire crew was awarded the Chief of Naval Operations Search and Rescue Model Manager Aviation SAR Crew of the Year for their selfless actions that day. The SAR Excellence Award is presented to recipients judged to exemplify the commitment to others embodied in the SAR motto: So Others May Live.

Some of the appeal of becoming a U.S. Navy aviation rescue swimmer is that you never know what you're going to be called to do on any given day. Even the mission-sets you brief and launch for will sometimes change mid-flight. Since the U.S. Navy helicopter aircrew are trained for so many different mission types, they may get diverted if they are the nearest SAR asset. This cliffside rescue had a unique medical emergency that does not appear in any military training manuals. They had never dealt with that exact scenario, but they adapted to the call of duty. At that point in his career, Worth had thousands of helicopter flight hours with a plethora of different mission-sets, to the point that he was numb to the adrenaline rush that comes when they heard the call to launch on a rescue. His and Highfill's meticulous attention to detail in an extremely disastrous and time-sensitive scenario helped them both remain calm and save a life. Training teaches the basics, but repetition and experience creates a more efficient and effective aircrewman. Special operations are special because they've learned the basics very well and have built on those foundational skills. That's what makes them a special asset to the military.

* * *

Psychology Today *defines meditation as "a mental exercise that trains attention and awareness. Its purpose is often to curb reactivity to one's negative thoughts and feelings, which, though they may be disturbing and upsetting and hijack attention from moment to moment, are invariably fleeting."*

Meditation can be used to enjoy the present moment and prevent distraction. A tool of meditation is mindfulness (discussed in chapter ten), which can involve turning your focus to your breathing and/or your five senses to bring more awareness. With meditation you can also choose to focus on a positive affirmation or scripture. When your mind wanders, bring it back to the practice you're focusing on to keep your mind in the moment. These practices are shown to decrease stress and anxiety and increase self-awareness and presence in the moment. Try it for a few minutes a day and increase time if desired.

* * *

During moments of chaos, it's easy to lose sight of the details and focus on the distracting elements of the problem. Focused breathing and slowing down helps us pay close attention to the small basic things that are easily overlooked but will cause the mission to fail or even cost lives. This holds true in everyone's unique world as we get caught up in the endless distractions and accelerating timelines of our busy lives. Taking a moment to be still, breathe, identify, and acknowledge the basic details helps paint a more complete picture. That picture then gives us a more thorough understanding of the situation to better prioritize our thoughts, emotions, and actions. When life becomes too arduous, take a moment to be still and breathe. That minor pause and breath may be all you need to find peace and clarity in that moment.

Chapter 7

Faith

On January 17, 1991, Operation Desert Storm launched with mammoth air strikes in Iraq and Kuwait from the United States and its allies. The first phase of the war was to achieve air superiority, take out air defenses, strike strategic Iraqi defenses, and obliterate Iraq's elite Republican Guard to encourage Iraq to relinquish control of Kuwait. After the first three days of successful bombing, the 3rd Marine Aircraft Wing flew around the clock sorties to carry out hits on specific targets in Kuwait. The Marines flew AV8-B Harriers and F/A-18 Hornets to destroy large quantities of air defense and artillery weapons.

Captain Michael Craig Berryman flew continuous air strikes with Marine Attack Squadron 311 (VMA-311), decimating Iraqi strategic weapons systems to help win the war. On January 28, Berryman was flying his AV-8B with a full complement of ordnance toward an enemy site in southern Kuwait. Then without warning, he experienced the violent and blinding impact of his Harrier being hit by an Iraqi surface-to-air missile. Due to an earlier aircraft malfunction, he wasn't able to detect the enemy missile until it was too late. The cockpit warning lights blared red with alarms ringing in Berryman's helmet. The damaged and disabled aircraft flipped upside down; with gravity pinning the pilot from negative g-forces, he desperately reached down for his ejection handle. With one forceful pull, the mild detonating cord activated and shattered the canopy as the ejection seat ignited and sent Berryman through the fragmented Plexiglas and acrylic. The Marine captain, strapped in his S-III-S seat, shot headfirst toward the hot desert below. Fortunately, his parachute automatically deployed and righted him prior to ground impact. (The S-III-S ejection seat is unique because it incorporates a sensor that is integrated with the aircraft pilot system that indicates aircraft air speed. The AV8–B can hover, so the ejection seat is designed for 0 to 600 knots.)

Once safely on the ground, Berryman diligently removed himself from the ejection seat, grabbed essential survival gear, and took off running. Shock and fear overwhelmed the young captain, but he never stopped running. He made it over a mile before Iraqi ground forces caught him and captured the Marine at gunpoint. He was taken back to an interrogation camp as an Iraqi POW. However, Berryman was never acknowledged as a POW by the Iraqis until after the war ended. Without knowledge of his whereabouts, he was listed as missing in action and presumed dead. For the next thirty-seven days, the Marine Corps captain would experience severe and continuous torture.

Berryman recalls that first day's encounter with Saddam Hussein's men in the prisoner camp. One man broke the fibula in his left leg by smashing a metal pipe against it. Another man repeatedly beat his right leg with an axe handle. The captured pilot recalls how much Saddam's special units and secret police enjoyed the cruelty of beating the Americans. They burned Berryman's forehead, nose, ears, and even an open neck wound with a lit cigarette. Despite the horrific treatment he was receiving, he recalls that the beatings weren't the worst part. The hardest thing was the isolation and not knowing what his family knew or was thinking of his situation. It crushed him to think of his family's agony not knowing if he was dead or alive.

His wife, Leigh, was a weapons tester for the U.S. Army when she heard the news that her husband had been shot down and was missing in action. She never gave up hope of his return, even though she was being fed damaging misinformation from the media. One day she received a surprising call from U.S. senator John McCain, which really helped her cope. McCain was taken prisoner of war in 1967 when his A-4E Skyhawk was shot down during his twenty-third bombing mission over North Vietnam. McCain sustained severe injuries from the ejection. When he was captured, the Vietnamese were going to let him die until they found out his father was serving as commander-in-chief of the

Pacific Command and spared his life. They tried to use him for propaganda purposes and offered to release him. McCain refused, following the Code of Conduct (CoC), and was sent to Ho Loa Prison (aka Hanoi Hilton). He spent five and a half years as a prisoner, with two of those excruciating years in solitary confinement. He was tortured beyond belief through his time in captivity but eventually survived the impossible and was released in 1973. An empathetic call from Senator McCain was the greatest gift of encouragement Leigh Berryman could ask for in that difficult time.

The first month in captivity, Berryman was completely alone, cut off from the other prisoners. The Iraqi interrogators insistently tortured him, trying to extract information about the U.S. intentions for an impending ground forces attack. The enemy military did not follow the Geneva Convention, which are the treaties and protocols established for international legal standards for humanitarian treatment in war. Berryman's beatings went far beyond torture, but he has honorably refused to publicly elaborate on the extent of his abuse to ensure the safety of those still in harm's way.

While in isolation, his prisoner camp was bombed by allied forces and he was forced to move to the Iraqi secret police headquarters, which they called Baghdad Biltmore. This camp was holding other American POWs, who Berryman joined. The prisoners recalled the unrelenting torture and misery they received at the Biltmore. Five agonizing weeks into the war, U.S. Air Force B-2 Stealth bombers dropped their payload on the location, destroying the Iraqi secret police headquarters. U.S. Intel and the B-2 pilots had no idea that American prisoners were being held there and almost took them out. Fortunately, they survived, and since it was evident the Iraqis were losing the war, they released the American heroes to the Baghdad Red Cross. After receiving medical care and being debriefed, they made contact with their families and were flown to freedom. Captain Berryman said if it wasn't for his strong faith in God and deep love of his family, he couldn't have survived the experience. Article 38, third paragraph of the Geneva Convention IV provides that protected persons "shall be allowed to practice their religion and to receive spiritual assistance from ministers of their faith."

Berryman commented in a 1991 interview with the *Oklahoman* newspaper, "You have a lot of time to reflect while you're in captivity. All of us, I think, became much more close with the Lord and in our religious beliefs. And I think it makes you more tolerant of the things that are going on around you. It helps you kind of realign your priorities as to what's important in your life and what's not."

* * *

From his survival experiences as a prisoner in a Nazi concentration camp, Viktor Frankl realized those that had hope and found meaning in their life survived where those that did not perished. "Those who have a 'why' to live can bear with almost any 'how,'" he said. Our faith and motivation to live can give us the courage and fight to keep going even in the most horrific circumstances.

* * *

On December 26, 2004, the third largest earthquake ever recorded struck off the west coast of northern Sumatra, Indonesia. The undersea megathrust earthquake known as the Sumatra-Andaman earthquake hit at a magnitude of 9.1 to 9.3, causing a 100-foot tsunami to impact fourteen different surrounding country coasts. The devastating natural disaster killed an estimated 227,898 people. The world responded to one of the largest humanitarian efforts in history. The USS *Abraham Lincoln* Nimitz-class aircraft carrier was on scene with its carrier strike group, including USS *Shoup*, USS *Shiloh*, USS *Benfold*, and USNS *Rainier*. The USS *Bonhomme Richard* was also in the region to provide aid along with its expeditionary strike group, including USS *Duluth*, USS *Milius*, USS *Rushmore*, USS *Bunker Hill*, USS *Thach*, and USCGC *Munro*. Additionally, U.S. maritime pre-positioning ships from Guam and Korea arrived stocked with food, fresh water, and relief supplies. Several countries and additional U.S. vessels would help in the weeks to come.

Helicopter squadrons HS-2, HSL-47, and HC-11 immediately began preparing for Humanitarian Aid / Disaster Relief (HA/DR) missions. They stripped their four SH-60F, SH-60B, and MH-60S helicopters of all anti-submarine equipment from the cabin area to maximize space for passengers and supplies. HS-2 had three HH-60H helicopters as well, which were already gutted for their CSAR primary mission. The aircraft carrier operated 15 miles off the northeast coast of Banda Aceh, Indonesia. All the fixed-wing aircraft were moved below to the hangar deck to allow maximum space on the flight deck for all the helicopters and supplies.

The squadrons had to get a waiver approved to support the necessary increased flight hours, aircraft configurations, and turnaround times to conduct daily missions into the disaster area. They would basically fly twelve-hour days from sunrise to sunset, eat, sleep, and repeat. AW2 Cory Merritt was an aircrew / aviation rescue swimmer early in his career. He recalls the days being like *Groundhog Day*, the 1993 blockbuster movie starring Bill Murray and Andie MacDowell where the characters repeat the same day over and over.

The squadrons filled the helicopter cabins with ships personnel to assist with distributing aid. They would then fly to a designated area to drop off the sailors, who would report to their aid stations. They then packed the helicopter cabin with supplies and the aircrew took off to deliver the aid. It was the Wild West as multiple aircraft swarmed the skies delivering supplies and searching for survivors. To those who had lost everything, the distant thumping sounds of the helicopter rotors restored a glimmer of hope and faith. Merritt recalls the total chaos of landing in an area to deliver food and water. Hundreds of starving and dehydrated victims rushed toward the dangerous helicopter. The aircrew made "pointy-talky" posters to try to communicate despite the language barriers. The posters would show the spinning helicopter rotors with heads getting cut off, but that didn't seem to faze most. They would rush into the spinning rotor arc and attempt to jump into the cabin or even through the port side window. The aircrew would hand out supplies as fast as they could and attempt crowd control, but with little effect. There were also terrorist groups firing off 7.62 mm rounds from their AK-47s right outside the rotor arc. The desperate victims would fight over the packages and even rip the boxes open, flinging the cardboard in the air to be shredded by the main rotors. It was a total hell zone of desperation.

The helicopters also searched for stranded survivors to extract and fly back to a triage site. Malnourished survivors were stuck for days on watercrafts, islands, and peninsulas created from the tsunami. Merritt remembers the complete devastation they flew over, with dead bodies floating everywhere. The grateful survivors were packed into the Sikorsky helicopters like sardines, maximizing all possible space. Once they dropped a load of survivors, the military and humanitarian medics would take over at the triage center to provide first aid. The helicopters were then filled up with supplies, refuel (if needed), and head back out.

One early morning on the flight deck of the USS *Abraham Lincoln*, five Navy helicopters waited for the sun to rise before heading out for the day's humanitarian missions. Most of the aircraft weren't on auxiliary power unit yet, which provides AC power for electrical, communication, navigation, flight control, and hydraulic ground checks before spinning up. An HSL-47 SH-60B was powered up with the rotors spinning on spot 3 of the flight deck. There are several designated spots on the flight deck for helicopters to land, refuel, and be chocked and chained for safety. AW2 Merritt was pre-flighting his helicopter at the far end of the carrier's flight deck about 300 yards from the SH-60B. Suddenly, a whistle blew through the flight deck public address system with the following message. "Man Overboard! Man Overboard! Port side! This is NOT a drill!" With Merritt's crew not ready to launch, he ran down the flight line to see which SAR assets were ready and if any help was needed.

In the HSL-47 helicopter, the aircraft commander, Commander Frank Michael, directed his crew to prepare for a SAR mission. Co-Pilot Lieutenant Robert Beeman and Crew Chief Gerard Schwarz changed their configuration from a passenger transfer to SAR ready. That's when Merritt arrived at the cabin door. Schwarz immediately said they needed him and to go get his SAR gear. Merritt ran the length of the flight deck back to his helicopter, grabbed his 50-pound bag of gear, and sprinted back to the HSL-47 helicopter. Merritt dressed out in his wetsuit, harness, mask, and fins as Commander Michael lifted the aircraft, turned, and searched for the man overboard.

Within a few minutes they spotted the survivor bobbing in the water a mile behind the aircraft carrier. As they flew a SAR pattern toward the struggling sailor, Schwarz slid open the cabin door and threw two MK-25 smokes to mark his position. Michael flew back around into the wind and dropped down to 10 feet / 10 knots. Schwarz readied Merritt who was sitting in the cabin door, and then tapped him three times on the shoulder. Merritt jumped from the SH-60B, gave the OK hand signal, and swam toward the survivor. The crew had an audience as the ship's company watched the entire event unfold on the Pilot Landing Aid Television in real time, which is unusual for rescues since they don't typically occur so close to the ship.

The man overboard removed his blue coveralls, tied off the feet, and filled them with air by slamming them down on the ocean surface to use them as flotation. Navy personnel are trained to use their clothing for buoyancy to save their lives in these types of situations. Merritt was surprised to see the man only in his underwear, drown-proofing to stay afloat, but didn't let it faze him or distract from his rescue procedures. The rescue swimmer conducted disentanglement techniques to ensure nothing was attached to the survivor, and as Merritt looked up to signal for the hoist, Schwarz was already lowering it to his position. Merritt connected the rescue strop around the submissive survivor, secured the safety strap around his chest, then connected his harness-lifting V-ring to the rescue hoist and gave the crew chief a thumbs up.

They were back onboard the *Lincoln* within minutes, where medics rushed in under the rotor arc to bring the survivor down to medical to be checked. It was a great example of the flexibility of the helicopter mission-sets across different squadrons and aircraft types. The standardization of search and rescue remains the same across the various platforms. Merritt laughed as he recalled walking down to the HS-2 ready room (briefing room) still in his rescue gear, holding his mask and fins, and dripping water. His aircrew chief took one look at him, threw him a mop, and told him to clean up the mess he was making. Merritt was awarded a Navy Achievement Medal for rescuing the man overboard.

A week later, Merritt was back in Indonesia. Lieutenant Commander Ruben Ramos and Lieutenant Junior Grade Bill Stickney piloted the SH-60F with Crew Chief Geoff Larson and AW2 Merritt in the back. They briefed for the day's HA/DR mission and loaded ten USS *Abraham Lincoln* sailors to assist with distributing aid. All aircraft gauges were green as they lifted from the aircraft carrier and headed toward the mainland. The record tsunami that had retracted the Indian Ocean over a mile out and returned with devastating force left the flight path to shore a mass casualty zone. The short flight was eerily quiet as everyone processed visions of what they could never unsee.

The Navy helicopter approached the landing zone, which was a soccer field designated for supplying aid. As the helicopter aircraft commander turned into the approach at 250 feet, the aircraft didn't respond to the flight controls for the turn. The SH-60F lost tail rotor effectiveness and the experienced pilot immediately reacted. The emergency procedure for a tail rotor drive failure is to autorotate to quickly get the aircraft on the ground using gravity and wind force to spin the rotors instead of using engine power. However, this was a tail rotor control failure rather than a tail rotor drive failure. Both failures can result in total catastrophe, but now have different emergency procedures. At the time they were handled the same. Ramos yelled "AUTO! AUTO! AUTO!" as he dropped the collective control and pinned the right foot pedal, attempting to stabilize the spinning aircraft. The aircrew only had a few seconds to go through their ditching procedures. Merritt had the right state of mind to close the cabin door, which most certainly saved the lives of the passengers.

What happened next was recorded for the world to see by the news media already on the scene, and the families of the aircrew later watched the horrific crash from the other side of the world. As Ramos pinned down the collective control and the crew did their best to get in a crash position, the seven-ton helicopter dropped straight from the sky down to a rice paddy. It managed to spin two and a half times at hard g-forces, pinning the two aircrew and ten passengers against the cabin walls. Merritt's nose exploded as he was crushed against the cabin door, and he ripped the muscles around his ribcage. If Merritt hadn't had the presence of mind to close the cabin door just seconds earlier, most or all of the cabin crew would have been ejected from the spinning aircraft. His quick actions saved the lives of everyone packed into the cabin of the crashing SH-60F.

The helicopter struts hit with punishing force, causing it to roll over on its starboard side. This caused a violent rotor strike, jolting the sideways aircraft up and down with indescribable force. The rotor blades exploded on impact, spraying debris across the field of water. The nearby humanitarian crews stared in disbelief as they processed what they were witnessing. It was that moment that everyone experienced their own fight, flight, or freeze. In most cases, like experiencing an earthquake or other natural disaster, people tend to freeze as they question and process what's occurring. *Is this really happening?* After a few moments, some of the bystanders unfroze from their gasping stares and ran to assist the crash victims. Others followed as well as local medical teams, who grabbed supplies to aid the injured passengers, not knowing if any of them could have possibly survived.

The ferocious crash seemed to last forever, but in reality, it was just a few seconds. But a few seconds that would replay in the nightmares of those on board for a lifetime. The SH-60F settled on its side, with the high-pitched sounds of the main engines and spinning rotor mast and hub, minus the rotors. All mechanical components eventually slowed down quieter and quieter, until they fell silent. Cries and moans filled the cabin as it lay on its side sinking a few feet into the field's muddy waters. Merritt was pinned underwater with Larson and ten passengers on top of him. He remained calm, as it was all he could do as Larson and the passengers began peeling off the pile to find a safe exit.

Many of the crew and passengers broke their backs and pelvises in addidition to sustaining other serious injuries. All suffered from shock. Merritt seemed to be the least injured and went straight to work performing advanced first-aid assessments. He recalls stabilizing one passenger but struggling because of the excessive amount of blood everywhere. He fervently searched for an arterial lesion but couldn't find the source. That's when the passenger told him it wasn't from him, but from Merritt. The rescue swimmer felt his demolished nose to find the source of the blood and grabbed a battle dressing to stop the flow.

The pilots were able to crawl out of the sideways helicopter, with the aircrew and passengers surfacing from the port side pilot's door one by one. Merritt was unable to pull himself free of the aircraft due to his rib injuries. He slowly inched his way forward to be assisted by the local helpers who had now made their way to the wreckage. All of the injured aircrew and passengers were treated locally with U.S., Canadian, and Australian medical support. They were then prepared to fly to the USNS

Mercy hospital ship for further medical treatment. Merritt climbed aboard an MH-60S, which was the newest model of Navy helicopter he had never flown in before. When they approached the USNS *Mercy* the pilots pulled a side flare maneuver to land on the ship's helicopter deck. Merritt had not experienced this new landing capability, which is a hard 90-degree swipe of the aircraft to land. CH-46 Sea Knight helicopters have conducted this maneuver for VERTREP for years, but it had just recently been introduced to the MH-60S community. Having just experienced a tail rotor failure crash, Merritt thought he was experiencing another. He had a moment of disbelief and uncertainty, but then looked out the cabin window to see the helicopter safely landing on the hospital ship's landing deck.

The crew was escorted down to the ship's triage area, where they received full assessments and treatment. They then flew back to the USS *Abraham Lincoln*, where they would get some rest and conduct a military mishap investigations. This includes drawing blood and taking urine samples to rule out drugs or alcohol in any of the aircrew, which could have affected their flight abilities, then a series of interviews to understand the day's events from all perspectives. The investigation is traumatic as it forces the aircrew and passengers to replay the recent horrific events, but it's necessary to understand and learn the details of the mishap to prevent them in the future. The military is always learning and evolving their technology and procedures to ensure they remain as effective and efficient as possible. Unfortunately, the lessons learned sometimes come from the losses of their heroes. And by losses, I don't only mean death. Mishaps and gruesome events many times lead to long-term diagnoses of PTSD, which can be a lifetime condition of flashbacks, nightmares, anxiety, and intrusive thoughts.

AW2 Cory Merritt rested on the aircraft carrier as HS-2 continued their daily mission-sets into Sumatra. Nine days later he rejoined his squadron on the long twelve-hour days of operation. Merritt earned the Navy and Marine Corps Commendation Medal for his quick life-saving actions during the SH-60F crash. He also earned another Commendation Medal for his humanitarian efforts in Sumatra. He completed a lucrative twenty-one-year career in the Navy, retiring as a senior chief petty officer in North Carolina. He and his wife, Jessica, started the Special Liberty Project, which serves veteran families, healing veterans, and Gold Star spouses and children. The Special Liberty Project provides a nature-centric and therapeutic environment, where the veteran families build the much-needed camaraderie and support that only veterans will understand.

* * *

Survival Evasion Resistance and Escape (SERE) training is arguably the most important training taught in any branch of the U.S. military. Its sole purpose is to prepare U.S. military personnel, U.S. Department of Defense (DOD) civilians, and private military contractors to survive if they find themselves shot down, lost, or taken captive in enemy territory. Lieutenant Commander Lou Conter, who survived the attacks in Pearl Harbor on December 7, 1941, was later shot down twice during World War II. He went on to fly over twenty-nine combat missions in the Korean War and would then establish the Navy's first SERE program in the 1950s. The original program, tailored to Navy pilots and aircrewmen, primarily focused on their survival if shot down. In a 2021 article, Conter told the *Navy Times*, "We had to teach them how to live in the jungle . . . how to hide in the jungle, how to navigate, and how to get out alive. Above all, don't panic in any situation you are given in life!"

All service members receive basic CoC training in boot camp, which is reviewed annually (level A). However, not all military personnel go through SERE training. Its main focus is on those with a high risk of capture (level C), such as aircrew, special forces, special operations, and anyone who could find themselves behind enemy lines and possibly become POWs. Each branch has their own version of training, based on their specific personnel and scenarios. The Navy and Marine Corps train together since the Marine Corps is a division of the Department of the Navy. SERE is brutal, but it's a game changer for developing the skills and confidence to honorably make it home alive. Some of the tactics and training are classified, so I'll share just a brief summary.

Currently, the Navy conducts SERE training at SERE West NASNI, SERE East Portsmouth Naval Yard, Marine Forces Special Operations

Command (MARSOC) Camp Lejeune, and Naval Special Warfare Naval Amphibious Base (NSW NAB) respective to the East and West Coast naval operations. Depending on the time of year, the East Coasters may have the opportunity to conduct their training in the freezing cold of winter. The West Coasters have the extreme desert weather, which is freezing at night and boiling hot during the day. No matter the weather, they all train to survive in harsh conditions with limited supplies. Although the course is taught by military and former military personnel, they also include former prisoners of war to provide real accounts of their experiences. It's extremely powerful to hear and learn from the heroes that survived the unthinkable.

The first week of the two-week course is conducted in the classroom, teaching academic lessons that are applied in the field during the second week. The SERE instructors teach the six articles of the CoC created after the Korean War, due to the number of POW deaths and prisoners not wanting to return to the United States because of the cruel North Korean psychological manipulation and torture. The CoC is a moral and legal ethics guideline for military service members on how to act if evading or captured in combat.

Code of Conduct

1. I am an American fighting in the forces which guard my country and our way of life. I am prepared to give my life in their defense.

2. I will never surrender of my own free will. If in command, I will never surrender the members of my command while they still have the means to resist.

3. If I am captured I will continue to resist by all means available. I will make every effort to escape and aid others to escape. I will accept neither parole nor special favors from the enemy.

4. If I become a prisoner of war, I will keep faith with my fellow prisoners. I will give no information nor take part in any action which might be harmful to my comrades. If I am senior, I will take

command. If not I will obey the lawful orders of those appointed over me and will back them up in every way.

5. When questioned, should I become a prisoner of war, I am required to give name, rank, service number, and date of birth. I will evade answering further questions to the utmost of my ability. I will make no oral or written statements disloyal to my country and its allies or harmful to their cause.

6. I will never forget that I am an American, fighting for freedom, responsible for my actions, and dedicated to the principles which made my country free. I will trust in my God and in the United States of America.

The classroom training includes in-depth details on all aspects of what the students will encounter the following week during their immersive field training. This ranges from wilderness and desert survival, psychology of survival, navigation (map reading, compass use, celestial navigation), foraging for and identifying food, developing water stills, building shelters, communication, signaling, basic medicine, camouflage, concealment, and rescue devices. The instructors engrain the four-step decision-making process of OODA Loop, which stands for Observe, Orient, Decide, and Act. OODA Loop allows the survivor to assess the existing information to make life-saving decisions and to reevaluate as more data becomes available. They teach the students about their body's ability to survive without food by meeting their basic need for water, which they will put to the test the following week when they have limited food based only on what they can find or what is provided. It also includes the mental and emotional aspect of survival. *If you lose your mind, you lose your life.*

SURVIVAL

A cool fall morning in 1994, during the second week of training, we were dressed in oversized worn, musty military clothing. The SERE instructors loaded everyone on a long, white school bus and drove a couple hours to a remote training site in Southern California. I remember the odd anxiety of wanting to get there to get it over with, but at the same time wanting the ride to take longer to avoid what was to come. As soon as we arrive, it was game on! We all gathered in a circle to receive basic instructions that aligned with our classroom survival knowledge. We were told it was time for some practical application of what we learned.

We used the natural resources of the desert to camouflage our faces and bodies. We built shelter with the limited objects we found from the surrounding landscape. We built water stills in the ground with small plastic bags, rocks, and vegetation. We used iodine pills to kill waterborne diseases like *Giardia* and cryptosporidium, to prevent explosive diarrhea and life-threatening illness. We built fires using a few different methods like rubbing sticks, hitting rocks, and eventually flint.

As a team, we caught and humanely ended a rabbit's life. We snapped its neck, skinned it from legs to ears, and boiled it for our last meal. That nasty bunny soup was shared by thirty hungry SERE candidates. We were taught every possible survival use of utilizing the rabbit's body, including food, rabbit hide, bones, intestines, and even the ears for eye protection against the blistering sun. Starvation is an area of training that can't be learned from a book. I had previously experienced land survival training in Pensacola, where we didn't eat for a few days while learning to survive off the land. (Land survival is no longer included in the Naval Aircrew Candidate School curriculum.) That particular training occurred during my nineteenth birthday. Happy Birthday to me! But a few days without food is quite different than a week, especially in our high-intensity training and youth, when our bodies were used to burning more calories than at any point in our lives. The human body can survive for three weeks without food, three days without water, and three minutes without air. However, the body will eat all its glucose and fat, leaving the person with severe, life-threatening symptoms.

The California desert temperatures fall rapidly at night since the lack of humidity makes the air unable to hold heat generated by the sand. We stuffed our clothing with leaves and moss to create some form of warmth to survive the night. Students are encouraged to utilize their body heat by remaining close together. We barely slept through the frigid night even being bundled together. Most of us experienced uncontrollable shivering and couldn't wait for the sun to rise, but that also meant we would be moving to the dreaded next phase of training.

Aviation Rescue Swimmer and SERE instructor for both SERE West and Naval Special Warfare (NSW) SERE, Whitney Warren, experienced an intense survival situation during her first few years in the Navy. During a higher-than-normal helicopter jump incident, she collapsed her lungs after impacting the water near terminal velocity. She recalls waking up in the University of California San Diego Medical Center and hearing the doctor tell her, "I don't know if you believe in God, but God believes in you!"

The medical professionals didn't think Warren would be able to fully recover and continue her hard-earned path as an aviation rescue swimmer. However, she beat the odds and miraculously recovered. She went on to a very successful military career with multiple deployments to the Persian Gulf, earning two Combat Air Medals for acts of heroism. With her near-death experience and years in the fleet, including as an instructor, she feels completely confident in SERE training. Warren says it teaches everything based on case studies from actual POWs, hostages, and Peacetime Governmental Detention. Faith is a core part of Warren's life as well as a critical element included in all areas of SERE training; to ensure captees return home with honor by surviving while protecting valuable information.

EVASION

The SERE instructors are spread all throughout the landscape, constantly in character. During my training in the 1990s, we were at the tail end of the Cold War, so the instructors held strong accents and carried AK-47s. They train and alter their dress and accents to mimic enemy soldiers in current conflict combat zones. There are safe houses we started from after being given instructions on how to navigate to checkpoints and another safe house miles away. There are no trails or roads, so we learned to navigate via topographic maps and avoid areas where we would be exposed and captured. We paired up with another student and worked together to try to reach our destination without being detected or captured within a specified timeframe. This meant crawling for most of the duration, remaining silent, and covering our tracks.

The artificial atmosphere of war the SERE instructors create is extremely believable. Off-road vehicles frequently drove through the area searching for any sign of human life. They shouted offensive comments through their megaphones to entice the students to give themselves up. I recall while hunkered down in the bushes on a side of a hill seeing in the distance others being captured and punished. At one point, an enemy soldier unloaded his AK-47 over my concealed hiding spot in the brush, raining down empty 7.62mm blank shells on my position. When he left, my partner and I remained still for a few minutes to ensure they weren't waiting for us. We then continued slowly crawling for another mile.

There were multiple evasion exercises, with added challenges and twists. We were heightened with fear and doing everything possible to remain unnoticed and uncaptured. If for some reason the students are found and captured, they experience a punishing and then are released to continue evading. My partner and I were luckily never captured in this portion of the training, but still experienced physical harm once we reached the safe house. Even the "friendly" instructors at the safe house couldn't be trusted.

Eventually throughout the program, everyone must be captured. It's a terrifying experience to know you're now a POW, even in a controlled training environment. The SERE instructors dressing the part drive around and yell over their megaphones that it's time to give up! One by one, with weapons drawn, they welcome each prisoner, and you can probably imagine how. They then place a musty cloth sack over each of the prisoner's heads and pile them into the back of a truck. The next stop, prison camp!

Resistance

In the fortified prison camp, the prisoners or "war criminals" are placed in small individual confined areas. They sit in two long lines facing each other. With the musty bags over our heads, we were put in an uncomfortable stress position and instructed to stay in that position at all times. We were given a bucket to urinate and defecate in. I guess it was gracious of them to not make us share one tin for all thirty of us. Staticky speakers with various psychological music, baby cries, water drops, or propaganda continually blared at top volume. There is no sleeping permitted during the day or night. With no food or sleep, it became extremely painful as our legs began to cramp. It is impossible to remain still. And the SERE instructors counted on it! Prisoner guards watched the boxes for any slight movement, then they'd pounce, dragging the war criminal out and punishing them.

The resistance tactics we learned in the classroom were put to the ultimate test during our stay at the prison camp. The guards randomly pulled a prisoner out of their box and brought them to a room across the dusty camp for interrogation. In the dark and cold rooms, they used different methods to try to extract valuable information from the SERE students. This could range from offering incentives, to telling lies in an attempt to turn prisoners against each other, or physical methods of coercion. The goal is to stay alive by providing as little information as possible. If the captive refuses to give up any information, then they are of no use to the interrogators and will execute them. The difference between SERE and the real world is they won't actually kill the captives. However, if the prisoners go too far without cooperating, then they could be coerced more harshly.

The SERE instructors also placed the captives in a small, cramped confinement, and locked the top. Some prisoners had their claustrophobic fears exposed while some, like me, fell asleep. Both reactions receive similar punishment when pulled from the box, which I can't include. They then start the process over. Similar to a tree that will break if bent too far. The tree can bend multiple times to the point of breaking, but if it continues to bend without giving way, it will eventually snap. Like trees, we were interrogated to the point of breaking, but then we'd give enough information to snap back to our upright position, to then be bent again. It's a brutal and vicious cycle, but it's effective.

In 1955, President Eisenhower established the CoC as a legal guideline for military personnel captured by hostile forces. Article VI of the CoC states, "I will never forget that I am an American, fighting for freedom, responsible for my actions, and dedicated to the principles which made my country free. I will trust in my God and in the United States of America." Faith is a crucial form of hope as the U.S. military will never stop looking for the heroes captured or missing in harm's way!

During his Navy career, aviation rescue swimmer and SERE instructor Mike Rogers was volunteered for the 59th Joint Task Force mission to locate Commander Richard Rich, who had been missing in action for over thirty years since the Vietnam War ended. Rich's F-4B Phantom launched from the USS *Hancock* to fly sorties an hour outside of Hanoi and was hit by a surface-to-air missile. Rich's wingman did not spot a parachute or hear any emergency radio signals as his jet crashed into a rice paddy. Rogers and his team flew to Vietnam and searched the soldier's last known coordinates. They spent days in the humid and dense environment, exploring the horrendous grounds that cost the lives of over a million people. Many such missions don't result in a find, but Rogers was fortunate to recover the hero's body and return him to his loved ones to help provide closure. There are no words that can provide peace for the loss of a life, but it's reassuring to know the United States will always continue their search to return their heroes home with honor.

Escape

In the cramped prison camp boxes, the captives communicate using Morse code and other means. Despite being hungry, tired, sore, cold, and scared, they attempt to coordinate an escape plan. Just like anything in the military, they must follow the chain of command within the prisoners. This adds a level of complexity since they are under constant watch with loud psychological music and propaganda playing. In my 1993 training during a prisoner yard walk, we created a diversion by starting a fight in one area of the camp. One of the prisoners tackled another and then they rolled around on the dusty, dirt ground. This distraction gained the guard's attention, which allowed one of our pre-designated prisoners to squeeze through a small opening near the front gate. The escape was successful! However, it was short-lived, as we all were punished, even though the coordinated escape worked as planned. Just like a real-life

POW camp, there would be no celebration for a well-orchestrated and executed escape plan.

Days felt like weeks. Due to our war crimes, we were told it would be days longer than we anticipated. Even knowing that probably wasn't true, doubt and crushed morale drained the prisoners. Then we heard distant gunfire and explosions! The prison camp was under attack as the guards ran confused until finally surrendering. The psycho music stopped blaring through the speakers and was replaced by an announcement that Navy SEALs had taken control of the camp. The U.S. flag was raised on a pole in the center of camp, replacing the enemy flag, as the loudspeakers played "The Star-Spangled Banner." Cautiously, the prisoners removed the sack cloth from their heads and slowly crawled out of their prison boxes. As the beautiful song rang through the camp, the proud emotion streamed down each face. There wasn't a dry eye in camp. We survived SERE training!

* * *

The Thanh Hoa Bridge, known as the "Dragon's Jaw," spans the Song Ma River a few miles northeast of Thanh Hoa, Vietnam. The bridge was a vital passage between the different regions of North Vietnam during the war with the United States. Reinforcements and supplies were transported across the bridge for the Viet Cong fighting in South Vietnam. On April 4, 1965, the United States led a massive seventy-nine aircraft strike on the bridge and surrounding location. Air Force fighter pilot Captain Carlyle Smith "Smitty" Harris, flying an F-105 Thunderchief as callsign Steel 3, dropped his payload on the eastern side of the Dragon. During the bombing run, his aircraft took a significant hit from an enemy 37mm shell, forcing Steel 3 to ditch his aircraft.

Smitty jettisoned the canopy and pulled his ejection handles. It all happened so fast that he ended up breaking his left shoulder while punching out. He didn't focus on the extreme pain since he was slowly deploying down in his parachute near a village outside of Thanh Hoa. The villagers awaited his arrival and then stripped him of his gear and tied him up. Captain Harris was brought to Hanoi where he would spend almost eight excruciating years as a POW in the Hoa Lo prison, aka "Hanoi Hilton." He, along with hundreds of other Americans, including Senator John McCain and George "Bud" Day, would suffer indescribable torture and abuse. Smitty introduced a tap code to other prisoners, which was used in World War II by POWs tapping on pipes to communicate. The tap sequences matched a matrix of the alphabet, so the Americans could spell messages to each other without their captors knowing.

Captain Harris took unbelievable risks to teach the code to other prisoners to establish a much-needed form of communication in their isolated world. The tap code essentially became their morale lifeline. After being tortured for days and returning completely shattered and demoralized, the first thing the beaten captives would hear is a faint series of taps spelling out GBU—God Bless You. Harris documented an early account of his experience: "When things were really bad, there was a hierarchy of beliefs without which we could not have survived. The first was a belief in God, then our country, our fellow POWs and our family and friends back home. We simply must not let them down. And we gained strength to prevail over a brutal enemy by our firm foundation in these beliefs." The communication and knowledge that they weren't alone provided necessary hope. It was that hope that helped many of them survive that horrific experience. The POW Tap Code is taught to this day in SERE training as a form of covert communication between prisoners.

A fellow inmate to Captain Harris, Bud Day, was considered the most decorated military officer since General Douglas MacArthur, earning the Medal of Honor and Air Force Cross. In 2013, forty years after his return from the Hanoi Hilton, he suffered complications from cancer of the esophagus. During his final days, Smitty visited him in the hospital. Due to Bud's frail condition, he wasn't able to communicate. Bud reached for Smitty's hand and slowly tapped the following sequence: two taps followed by a pause and then two more taps signaled "G." One tap, a pause and two taps were a "B." Four taps, a pause and five taps meant "U." God Bless You.

* * *

There have been many findings about the benefits of faith on mental health. The National Alliance on Mental Illness (NAMI) states: "[S]pirituality is a sense of connections to something bigger than ourselves—it helps a person look within and understand themselves while also figuring out the greater answer of how they fit into the rest of the world. In other words, it helps people understand their interpretation of the meaning of life. Spirituality also incorporates healthy practices for the mind and body, which positively influences mental health and well-being. Here are some of those benefits: individuality, mindfulness, and unity with surroundings."

I feel that having a religious community helps in feeling connected and building a support system of people to help no matter the circumstances. Other research has also shown that spirituality has been protective against depressive symptoms and helps prevent experiencing burnout. On a personal note, I know that my faith has helped me get through many obstacles in life, including my experience while my husband, Brian, was climbing Mount Everest. He soloed the summit and then went completely snow-blind and witnessed a miracle from God on his descent. I experienced my own miracle back at home while praying for peace. Since I've always dealt with anxiety, I felt like God gave me peace in knowing that whatever might come, I have my salvation in Jesus and that God would take care of me and our two young children. Thankfully that wasn't the outcome, but I still learned a lot from the experience and the power of prayer and faith!

* * *

SERE training is unforgettably brutal, but I feel honored to have experienced it. Nobody comes out of SERE the same person they entered. SERE is physically and mentally challenging, and the students will learn something important about their strengths and weaknesses. Besides the advanced survival skills, the candidates learn their personal breaking points. Every SERE student will fail at some point and that's OK. SERE is one of the few schools where mistakes are encouraged. That's the point since we learn more from our failures than our successes. During their post-mortem interview from the SERE instructors, they will be faced with and learn from the humiliation they endured. Their eyes will be opened to mistakes and lessons learned, based on all the unique areas and tests that are intertwined in the program. Faith is drilled into the candidates, as without it there is no reason to survive. The enemy can break the body, but the body will be restored. Unrelenting faith provides a purpose beyond this torturous world and can help get us through the most indescribable and impossible times.

Chapter 8

Lean into Adversity

Retired U.S. Navy aviation rescue swimmer AWSCS Shawn Porter served thirty years primarily focused on CSAR operations. In the early 1990s, he deployed to Al Jowf, Saudi Arabia, during Operation Desert Storm with Helicopter Combat Support Special Squadron 5 (HCS-5), previously Helicopter Attack Squadron (Light) 5 (HA(L)-5). The Navy squadron was assigned to a joint military taskforce for security and CSAR. HCS-5 supported a National Guard A-10 Thunderbolt II "Warthog" strike force, which continually rained down hell on Iraqi targets with its fearsome nose-mounted GAU-8 30mm Gatling guns. The A-10 was credited with destroying 987 tanks, 926 artillery pieces, 1,355 combat vehicles, ten fighters on the ground, and two helicopters in an air-to-air engagement.

Early in his career and being the new guy, plus having his expert M60 aerial and ground qualification, landed Porter as security detail for the colonel. Rather than flying around-the-clock mission-sets, he was mostly down at sea-level riding in a Humvee. Many of his security tasks were during the night to reduce the enemy's ability to track their high-value individual's movements. Porter recalls wearing heavy body armor and the older night vision goggles on his helmet while manning the exposed roof-mounted M60 belt-fed machine gun. He experienced regular anxiety-heightened assignments as they securely transported the colonel and his party to strategic locations throughout the war.

The U.S. Air Force F-111 Aardvark and Raven played an important role during the Gulf War. Although it would be the aircraft's last major deployment, replaced by the F-15E Strike Eagles, the F-111 flew over five thousand missions jamming air-defense radars and dropping laser-guided bombs on Iraqi air defense, military installations, and ground forces. Unfortunately, an EF-111 crashed during evasive maneuvers, killing both pilots upon impact. Porter recalls his squadron launching for the SAR mission, which quickly turned into a body recovery for the lost aircrew. The emotional pivot is real in this line of work, as the mood of an aircrew can swivel from amped for the ability to save lives to the grim reality of the dangers, trauma, and sadness they often must deal with.

On March 31, 1998, Porter was launched from Naval Auxiliary Landing Field San Clemente Island located about 80 miles west of San Diego. An S-3B Viking was operating in the area when it experienced an engine failure and was forced to ditch into the shark-infested Pacific Ocean. The two pilots and two aircrewmen ejected safely and peacefully floated down to the frigid dangers below. Helicopter Combat Support Squadron 85 (HC-85) was first on scene. Porter, wearing a full wetsuit due to the cold-water temps, deployed from the UH-3H Sea King to rescue both pilots. Meanwhile, HC-11 was in the area as an additional SAR asset. A rescue swimmer jumped from their CH-46 Sea Knight to extract the S-3's remaining two aircrewmen.

On February 11, 2003, Porter was the wing SAR evaluator helping HC-85 out, as they were short-handed on aircraft and aircrew. During a routine nighttime starboard D flight with the aircraft carrier USS *John C. Stennis*, an EA-6B Prowler missed the arresting wires on landing and then experienced a brake failure. The aircraft passed across the flight deck and continued over the front of the carrier's angle deck. The pilots punched out as the Prowler disappeared into the darkness over the front of the ship. The carrier's air boss immediately sounded the alarm, and the massive warship turned hard to the starboard to prevent crushing the crew. The first pilot's chute barely had time to deploy before landing on the aircraft carrier's catwalk, where rescue crews rushed to his aid. The other pilot floated down toward the deck, but right before landing was dragged overboard by his parachute. The pilot fell swiftly seventy feet to the port side of the ship and entered the ocean. The HC-85 UH-3H was on top of the pilot within minutes, where Porter lowered down Rescue Swimmer JP Garcia. EA-6B Prowlers typically have four crew members, but there were only two visible parachutes descending after the crash. The HC-85 pilots had to confirm multiple times with the carrier that there were only two pilots on that flight, as they didn't want to leave the scene without conducting a thorough search if there were missing personnel. Due to excessive saltwater spraying in the cabin from the rotor wash, the ICS was malfunctioning. Porter recalls yelling to the pilots above the helicopter noise to transfer controls to the cabin crew hover so he could conduct the hoist operations. Crew chiefs train and are tested often for loss of communication and equipment malfunction procedures for this very reason. Once the ejected pilot was recovered, Porter closed the cabin door and proceeded with a medical assessment and treatment for shock. Even with the ICS issues, he recalls both rescues being pretty textbook. He says it was actually easier compared to the intensity of training. Shawn Porter was awarded separate Navy and Marine Corps Commendation Medals for both rescues.

Porter continued with multiple deployments to the Persian Gulf, and due to his combat experience and role, had his security clearance upgraded to Top Secret / Sensitive Compartmented Information (TS/SCI). He deployed and conducted operations with Naval Special Warfare (NSW) and allied forces at all corners of the world. During Operation Iraqi Freedom, Joint Special Operations Task Force controlled the 160th Special Operations Aviation Regiment (SOAR), Navy Combat Search and Rescue, C130 Hercules aircraft, and all special forces. Porter, attached to Helicopter Combat Support Special Squadron 5 (HCS-5), was deployed to hot zones inside Iraq for strike rescue and NSW support. His squadron conducted over three hundred combat missions. Each mission-set exemplified leaning into adversity as each situation was meticulously planned out with limited intelligence, requiring critical adaptation in real time. HCS-5 was tasked with direct action inserting, extracting, and supporting Navy SEALs, Special Forces Operational Detachment Alpha, Explosive Ordnance Disposal (EOD) and Marine Forces Special Operations Command (MARSOC). Most of their missions were taking down houses, streets, or city blocks to search for high-value targets

(HVTs). During snatch-and-grab mission-sets, the Navy helicopters were equipped with GAU-17/A "miniguns" while they flew overhead to identify and discourage squirters. These were typically HVTs running from their hiding locations, at which point Porter's team would communicate their location to ground forces so they could be captured or killed.

Porter recalls a hairy situation in Iraq, where two HH-60H helicopters came in hot to hover over a building. They tossed out the fast rope and started deploying special forces on the rooftop. The hurricane force of the rotor wash ripped up a random carpet from the roof. They watched in almost slow motion as it circulated up into the air, barely missing the main rotors. The CSAR teams were accustomed to taking live fire from enemy insurgents, but they hadn't trained for dodging magic carpets in Iraq. However, they were committed to the mission, even with operators literally hanging by a thread. The foreign object could have easily wrapped around the rotors, causing the helicopter to crash. Fortunately, sometimes luck works out in your favor and the carpet missed the helicopter and flew over the side of the building. Porter and team avoided a catastrophic mishap and successfully completed the mission.

Porter credits his ability to remain calm in chaotic situations to his Vietnam War predecessors who later became instructors. Based on their real-world experiences, they were able to articulate proper procedures to conduct techniques and execute a mission. They taught to decipher and prioritize what was important and also included things not published in the training manuals. Some procedures were good to know, but they would never actually have to execute themselves. Porter served at a time when he and his peers had the opportunity to reshape Navy Special Warfare CSAR. They learned from the experienced experts of the Air Force and Army 160th. And they continue to learn and evolve to always improve their ability to maintain a competitive advantage to defend this great nation. A fun fact about Shawn Porter: He can be seen lighting up a truck with an M240 from an MH-60S during an epic nighttime fight scene in the 2012 blockbuster film, *Act of Valor*.

* * *

Adversity builds resilience. The following characteristics that resilient people possess are defined by verywellmind.com: A survivor mentality is when even though things are hard, you know you can get through the tough times. Emotional regulation is the ability to recognize and to manage emotions, even during high stress. Feeling in control is recognizing that your thoughts and actions play a role in the outcome. Self-compassion means treating yourself with kindness, self-acceptance, and self-love, especially when things are difficult. Problem-solving skills is looking at the problem rationally to develop a solution. Social support is having a healthy support system and knowing when you need to ask for help. All of these characteristics can be worked on individually or with a therapist to better develop resilience.

* * *

On a blistering hot and moonless August night in 1999, the USS *Constellation* and its battlegroup of ships and Carrier Air Wing Two entered the Persian Gulf. They would spend the next ten weeks conducting over five thousand sorties, with over twelve hundred of those in response to protecting no fly zones for Operation Southern Watch. Helicopter Anti-Submarine Squadron 2 (HS-2) was back in action, supporting the aircraft carrier and flying daily mission-sets throughout the Gulf. The CSAR crews flew countless hours of sea surface control and surface anti-surface warfare. The squadron was equipped with six SH-60F helicopters for SAR and anti-submarine warfare and two HH-60H helicopters, gutted, and configured for CSAR operations.

AW3 Ed Majcina was launched on a helicopter visit, board, search, and seizure (HVBSS) to take down an evasive Iraqi vessel operating near the Iranian border. Majcina was a part of two HH-60H crews (Dash One and Dash Two) inserting a SEAL team via fast rope and providing aerial gunner support while the SEALs took action on the vessel. The suspicious Iraqi cargo ship was steaming fast due north to evade the U.S. Navy. The situation became critical during the insertion as the ship's crew pushed full throttle and then locked the armored doors to the helm control room. They then fled down below deck and locked themselves in the engine room. The end result could be disastrous as with nobody in control, the ship would eventually crash into the shores of Iran.

The Navy helicopter pilots aggressively flew in formation to intercept the runaway enemy ship as the SEAL teams prepared for the HVBSS. Once on scene, the Dash One pilots matched the ship's speed as they hovered over the forward deck, while Dash Two held a secure pattern providing M240 aerial support. Dash One ran into issues trying to deploy their load of SEALs due to a fouled deck and quickly moved to the overwatch position allowing Dash Two to move in for SEAL insertion. (Overwatch is a force protection tactic in modern warfare where one small military unit, vehicle, or aircraft supports another friendly unit while the latter is executing fire and movement tactics.) Majcina ripped open the cabin door, connected the fast rope to the insertion/extraction system (known as a FRIES bar), and tossed out the 40mm rope down a narrow walkway. One by one, carrying M4 carbines, breaching equipment, and various tactical weapons, the SEAL team fast roped down as the helicopter held its position. Once the first CSAR helicopter emptied the special forces team, Majcina released the rope, and the aircrew manned their port and starboard-mounted M240s and replaced Dash One for aerial support overwatch. Dash One then moved back in to position to deploy the remaining SEAL operators. On the ship's deck the SEAL team fanned out with weapons drawn as they secured the area. As the final SEAL fast roped from the hovering helicopter, the ship suddenly jolted starboard, which shifted the team member's landing area. The special forces operator came off the end of the rope, but instead of landing on the ship's deck, he went over the side of a superstructure, forcefully landing down below. A SEAL medic attended to his injuries as the rest of the team continued with their mission to secure the cargo ship.

The Iraqi boat crew was prepared for a possible boarding as they made sure the wheelhouse was armored and locked. The helicopters remained overhead, flying in a defensive pattern as the cargo ship continued its path. They were operating so close to the Iranian shore that Majcina recalls being able to read license plates from nearby cars. The SEAL team had to work fast, as they were getting dangerously close to shallower Iranian waters. The armored door was a problem, but the teams solved it with the use of door-breaching explosives and a diamond saw to gain access. Once the door was breached, they tossed in flash grenades and cleared all sectors of the control room. With an empty room, they were able to quickly gain control of the ship and change its course back toward international waters.

During the overwatch, Dash One and Dash Two received radio threats from Iranian gunboats pursuing the two HH-60Hs. The two CSAR helicopters then heard the terrifying sound of their AN/AAR-47 Missile Approach Warning System detecting a missile lock from the Iranians. The Navy helicopters took evasive actions, with countermeasures ready to deploy in response to the launch of the Iranian surface-to-air weapons. With heightened anxiety, Dash One and Dash Two continued their mission as they were told to ignore the Iranian warnings since it was most likely an empty threat. That didn't help the crews rest easy as the radio chatter increased with the intensity of the unfolding situation.

Back onboard, a couple SEALs remained in the bridge to drive the boat, while a team went and cleared the rest of the ship. They forced their way into the engine room and zip tied the Iraqi prisoners. Dash One and Dash Two were then radioed in to recover some of the SEALs. Since there wasn't a safe area to land, they dropped a rope ladder for the SEALs to climb into the hovering HH-60H. The Iranian gunboats stayed in the area during the extraction, but then left the scene after the cargo ship was in U.S. control and back in international waters. A rigid hull inflatable boat was launched from a nearby U.S. Navy frigate, which extracted the injured SEAL, while the others stayed onboard with their Iraqi prisoners to drive the cargo ship to safety. Dash One and Dash Two returned to the USS *Constellation* to debrief the day's events and lessons learned. They were operating on the edge in a dangerous location with enemy operatives engaging, but their professionalism and calm allowed them to successfully complete the mission.

Majcina was honorably discharged from the Navy after serving eight years, with an impressive and arduous six deployments. He then spent a year working for Blackwater Security Company, providing protective services in Iraq. Majcina recalls a defining mission that changed the trajectory in his career. In 2004, he and his Blackwater operators were holed up in two fortified villas on the outskirts of Dohuk, nestled up against the Bessre Mountain Range. A U.S. State Department official received information of an imminent development occurring in Dohuk, which was quickly relayed to Peter M. Thompson, deputy governorate coordinator. Thompson knew of the Blackwater presence in the area and reached out to Majcina to discuss the situation.

Latifa Ali, an Iraqi-born citizen, had fled with her family to Australia after Saddam Hussein threatened to hunt and kill them. Ali was betrayed by her father when he tricked her into returning to the war-torn country and burned all her documents and passport. She was held captive by her family, who prevented her from accessing the Australian Consulate. She was imprisoned, violated, and abused as she was forced to work as a spy. She was eventually allowed to translate documents for the U.S. contingent in Dohuk, which brought money in for her family. Defying her orders to become a traditional Muslim Iraqi woman with an arranged marriage, her family lost patience and planned out her execution. She would be stoned to death. She knew she needed to attempt to escape her captives and that she would most likely die trying, but it was better than the alternative of remaining a prisoner.

After Blackwater mission commander Majcina received the details of Ali's imminent death, he went right to work preparing a rescue effort. At the time, this area of Iraq had minimal U.S. military presence, which meant they could not rely on additional support. He researched the intelligence he was presented and found patterns of Ali's daily commute as she walked down the same section of road each morning and night. Majcina put a quick response taskforce together to snatch and grab the Muslim captive during complete daylight, which wasn't their normal time of operation for such a mission.

Two armored trucks with mounted .50-caliber machine guns, packed with fully armed Blackwater operators, vigilantly made their way into downtown Dohuk. As soon as they reached the section of road Ali would be walking, Majcina immediately spotted her and jumped out of the truck. With a strapped Rhodesian ammo carrier, M4 carbine assault rifle, Glock 19, and frag grenades, he made his way to the person of interest and established communication. Ali was startled and visibly shaking when Majcina grabbed her arm and turned her around. The public disturbance quickly gained the attention of others in the area, who went straight to their cellphones to text and call local authorities. The Blackwater operator explained who he was and his intentions to the fearful Ali and then swiftly escorted her toward the trucks. Majcina had to forcefully navigate through the gathering crowd, using the barrel of his M4 to facilitate.

They reached the armored truck and placed Ali in the backseat wedged between two armed Blackwater men. Both vehicles sped out of the downtown area and onto Highway 2, which leads to Mosul. Ali, still shaking, was read a repatriation document by the State Department official. Majcina recalls what a beautiful moment it was to hear and see the reaction to the official letter of Ali's return to her country of citizenship. She wept in shock as she processed the moment. The Blackwater team drove the armored trucks the 80 kilometers to Mosul, running multiple armed checkpoints, fortunately without incident. The heightened mood and urgency were chaotic, but the team eventually arrived safely at Mosul International Airport, where they released the frightened woman into custody with their government contact. She was flown back to Australia, where she remains today.

Like many of us in the aviation rescue swimmer community, Majcina was driven and wired to run toward the fight rather than away. But we all have that moment where we wonder why the hell we're doing what we're doing. Majcina was at this crossroads, as bullets were flying, and he was constantly operating close to the edge trying to make a difference. Soon after he would hear of Latifa Ali's safe return and received a surprising call from Australia's prime minister, John Howard. Prime Minister Howard wanted to personally thank Majcina for his heroic planning and execution in returning one of their citizens. After that call and a Letter of Commendation from Peter M. Thompson, deputy governorate coordinator, he realized his purpose was to be there to help save Ali. He finally felt at peace with his decisions and decided to change his career. Ed Majcina has spent the last two decades as head of personal security for several well-known television shows and individuals. His list includes CBS's *Amazing Race*, where he kept the film crew and teams racing through various cities across the globe secure. Additionally, he got to test all the extreme obstacles and stunts before they let the contestants try them. His personal security portfolio has included Steve Forbes, Lady Gaga, Miley Cyrus, Ozzy Osbourne, Tool, Demi Lovato, Linkin Park, and Robert De Niro.

* * *

CSAR are overland and overwater SAR operations occurring near or in combat zones. CSAR missions have been conducted since World War I and have significantly advanced over the past century with lessons learned and evolving aircraft, weapons, and tactics. Aviation rescue swimmers have the opportunity to develop their skills to fly CSAR missions with their respective squadron or receive orders to a dedicated CSAR squadron like HSC-85.

Navy CSAR mission-sets cover the gamut of providing special operations support and saving lives in combat zones. The pilots are specially trained to fly in mountainous ranges, conduct expedited, and evasive insertions and extractions of U.S. and joint forces while under enemy fire. Their training involves various tactics and maneuvers with other helicopters, fixed-wing support, and allied forces. The CSAR helicopters are equipped with specialized firepower to discourage enemy forces, while seamlessly conducting missions in hot zones.

During rising tensions in 2023 from the Israel war on Gaza, Iran-backed Houthi rebels from Yemen launched multiple attacks on commercial vessels entering the Red Sea. In response, the United States deployed multiple warships for Operation Prosperity Guardian to increase security in the critical global shipping lane. December 30 would end with a test of the U.S. Navy's helicopter aerial protection as the Singapore-based Maersk *Hangzhou* came under attack from four Houthi small boats. The USS *Dwight D. Eisenhower* and USS *Gravely* responded to the distress call. The aircraft carrier *Eisenhower* launched MH-60S helicopters from squadron HSC-7 Dusty Dogs in response to the attack. Once on scene, the Houthi boats fired small arms at the helicopters. The Navy door gunners returned .50-caliber GAU-21 automatic fire in self-defense, decimating three of the boats and killing ten Houthi rebels. After witnessing the overwhelming firepower of the U.S. rotor-winged units, the fourth boat fled from the scene.

The U.S. Army's 160th Special Operations Aviation Regiment (SOAR) "Night Stalkers" is the military's primary aviation unit supporting special operations. First established in 1981, a year after President Carter requested a unit be developed due to the failed attempt to rescue American hostages during Operation Eagle Claw in Tehran, Iran, the 160th Aviation Battalion saw its first combat in 1983's Operation Urgent Fury during the U.S. invasion of Grenada. They have since continued to evolve their intense mission-set and training to provide helicopter aviation support for special operations. SOAR provides support in attack, assault, and reconnaissance at night, at high speeds, low altitudes, and with short notice. They have contributed to the success of such high-profile missions as Mogadishu (depicted in the 2001 movie *Black Hawk Down*), rescue and recovery of PFC Jessica Lynch, the capture of Saddam Hussein during Operation Red Dawn, and the raid on Osama bin Laden's compound in Pakistan.

The U.S. Navy SEALs often fly covert operations with the 160th SOAR unit as well as with the U.S. Navy's dedicated squadron for special operations: Helicopter Sea Combat Squadron 85 (HSC-85). The combat squadron traces its roots to 1948 with the Navy's first operational helicopter squadron, Helicopter Unit Squadron 1 (HU-1). The mission-set and demand for rotary-wing assets grew significantly during the Vietnam War, when HU-1 was redesignated as Helicopter Combat Squadron 1 (HC-1) and broken up into four new squadrons. One of the new squadrons was the renowned Seawolves of Helicopter Attack Squadron (Light) 3 (HA(L)-3), which was primarily focused on special operations support.

During the Vietnam War, these specialized squadrons honed their procedures and tactics to remain effective in the unique jungle warfare and mission-sets. After the war, the Navy recognized the need to continue the focus and experience of the combat aviators and maintenance crews of the combat and attack rotary squadrons. They established Helicopter Wing Reserve (HELWINGRES) in 1975, located in NAS North Island, California, to continue the mission of HC-7 and HA(L)-3. A couple decades later, Helicopter Anti-Submarine Squadron 85 (HS-85) became part of HELWINGRES to take on the mission of search and rescue.

HS-85 continued their primary focus of search and rescue for another decade until they gained the designation for "combat" and became HSC-85. In 2011, the United States Special Operations Command (USSOCOM) asked the Navy to have a dedicated squadron to support the growing demand for special operations. HSC-85 was assigned this role with the name Firehawks; they use purpose-built HH-60H and MH-60S Sikorsky helicopters to support their world combat missions. HSC-85 is the only squadron dedicated to providing training and readiness support to Naval Special Warfare and other special units. Their highly skilled aircrew and maintenance crews deploy worldwide to provide top-tier support to the United States's military special operations.

Aviation Rescue Swimmer AWSCS Jason Barney (retired) spent the latter part of his career dedicated to CSAR. Not just as an operator, but as an instructor, finishing his career as an aircrew lead chief petty officer at HSC-85. He was tasked with developing the program to a higher standard, based on lessons learned and the evolving landscape of war and terrorism. This came from over two decades of experience being forward deployed to war zones, supporting special forces, and losing several close friends and colleagues along the way. Barney channeled this experience and loss into motivation to develop improvements and proficiencies for future CSAR operators. This includes all aspects required to make an effective and healthy operator: physically, mentally, and emotionally.

The CSAR aircrew train excessively in air, land, and sea navigation and planning along with the pilots and other supporting personnel. Detailed preparation is critical to the success of the missions, including contingency plans as things rarely go as expected. They train hard to build the procedural memory required to efficiently and effectively perform high-risk missions in dangerous environments. Procedural memories are implicit memories that form without effort. These memories are drilled into aviation rescue swimmers all throughout their water, land, and combat training to ensure they operate at the highest level. They utilize the provided intel to create a plan based on a predicted outcome with multiple contingency plans. Timing is critical since situations change quickly and coordination between the survivor and those involved in the rescue can be a moving target. As the direct point of contact, they need to have keen situational awareness and continually communicate the plan with the real-time changes. In most cases, they try to preserve the element of surprise to get in and out before the enemy knows they were there. But in the real world that doesn't always occur, and they are well prepared to respond to enemy resistance.

CSAR receive extensive weapons training, including the use of mounted machine guns and weapons carried when exiting the helicopter, to conduct the overland portion of a mission. They train in special tactics performed while extracting friendly survivors while taking enemy fire. This includes radio communication and interrogation of the survivor to ensure they don't pick up an armed, decoy enemy.

CSAR helicopters operate in pairs for support with alternating air cover as the rescue helicopters move in for insertions and extractions. CSAR personnel work to rapidly deploy special forces and special operations in any environmental situation required. The fastest method of deployment and extraction is to land the aircraft, but that's not always a possibility based on the topology, watercraft, weather, enemy activity, or urgency. The teams have specialized rope master training and equipment to rig the helicopter for rapid insertion and extraction methods like fast roping, rappelling, and special patrol insertion / extraction (SPIE). Fast roping is a technique used to quickly insert military forces from the helicopter by sliding down a thick rope. This is the fastest deployment method other than landing, but dangerous since the troops aren't attached to the rope other than with the own strength of their hands and feet. Rappelling is safer than fast roping since the troops are connected to the rope via a carabiner and rappelling device attached to their harness. However, it's not as quick since it takes more time to connect and disconnect the carabiner and rappelling device from the rope. And you can't have more than one person connected to the rope at a time, where fast roping can have multiple people sliding simultaneously. SPIE was developed for inserting and extracting special operations in areas where helicopters can't land. The CSAR helicopter hovers high above the extraction zone with a special line dropped to the operators below. Each member connects a carabiner attached to their harness to evenly spaced lifting V-rings on the SPIE rig. The helicopter then vertically lifts to clear the obstacles and short hauls the special operators to a safe area to land to board them so they can return to base.

Training occurs during both day and night, as you never know when you'll get the call for a CSAR mission. The aircrew utilizes night vision goggles (NVG) and the helicopters are equipped with specialized night technologies like forward-looking infrared (FLIR) cameras. The crews conduct exercises and evaluations in dangerous mountain ranges using NVGs during moonlight, starlight, and pitch darkness to gain knowledge and confidence in all night conditions. Combat training occurs in the mountains or desert beyond the military bases, on naval or merchant ships, and at the U.S. Navy's premier air-to-air and air-to-ground facility at Naval Air Station Fallon, Nevada. The Naval Fighter Weapons School, also known as Top Gun, was relocated to NAS Fallon in 1996, from its previous location at NAS Miramar, California. The desert mountains and seasonal extreme temps provide a perfect environment for coordinated military exercises, including all types of different aircraft, special warfare commands, and allied joint force commands.

CSAR training and operations require keen situational awareness for the safety and success of missions. With hundreds of moving parts, coordinated efforts, human involvement, and constant distractions, any one misstep can adversely affect the outcome of the mission. Planning, communicating, and constantly being aware of the micro and macro elements of the tasks is everyone's responsibility. The Navy may deploy a couple helicopters involved for combat support and special operations insertion and extraction, but they aren't working alone. There is bomber air support, drone operators, U.S. and allied ground forces, plus command and control for communications. The CSAR teams lean into the fight but must be as prepared as possible to respond to the unpredictable and everchanging scene of the mission.

In 1997, during my second deployment in the Persian Gulf, my helicopter squadron was working with a Navy EOD unit attached to the aircraft carrier. We were developing a fast deployment of a RHIB near the coast of Kuwait. A RHIB is a high-performance multi-chamber inflatable boat containing a rigid hull bottom with an outboard engine used for special operations stealth insertions. We called the operation

"Tethered Duck" as we developed a method to mount, transport, and release the partially inflated RHIB from the bottom of the helicopter. (This was later designated Kangaroo Duck or K-Duck, because the attached raft appears similar to a kangaroo's pouch and the boats are known as ducks in the rescue community.) During previous helicopter boat insertions, the deflated raft took up valuable space in the cabin, which reduced the number of rescue personnel and special operators. Tethered Duck would free up the cabin for more equipment and personnel, plus provide a more efficient and effective response to various scenarios. However, the configuration was a lot more complex than it originally sounded, as any uncoordinated release could send the inflated boat or straps into the spinning rotors or engine intake and cause a major mishap. Our teams also had to work quickly in a high-stress environment to strap the boat underneath the HH-60H in between flight operations on the aircraft carrier. The same new method of RHIB deployment was being tested with more focus and detail at Air Test and Evaluation Squadron VX-1 in Patuxent River, Maryland. VX-1's mission is to test and evaluate anti-submarine warfare (ASW) and maritime anti-surface warfare (SUW) weapon systems, airborne strategic weapon systems, support systems, equipment, and materials in an operational environment.

The RHIB was tightly secured with webbing straps under the helicopter, attached to a central cargo hook designed for VERTREP. The cabin was filled with five EOD operators and four aircrew. Another HH-60H containing the same complement of aircrew and EOD, minus a RHIB, trailed the lead helicopter. They took off from the USS *Constellation* and headed west toward Kuwait, maintaining a low speed due to the raft's additional wind drag. The CSAR aircraft had its main landing gear struts fully extended to allow for space to secure the boat, which didn't allow for much wiggle room in case of a mishap. The aircrew watched intently and physically held the straps to ensure the RHIB remained stable, with the crew chief and pilots ready to release the load in case of an incident. In an undisclosed location near the coast, the helicopter came into a slow and low forward hover. The pilots cleared the aircrew to release the raft as the crew chief and three aircrew simultaneously released their attachment points, which funneled back to the cargo hook.

As one released, the pressure of the boat weight caused the others to jam on the hook. The crew chief alerted the pilots and without hesitation, kicked loose the cargo hook lock, releasing the jammed strap and the RHIB dropped down to the Arabian Sea. With the raft deployed and the HH-60H clear, the crew chief moved to the open cabin door and deployed the EOD operators one by one. The second helicopter then came in behind and deployed their payload of EOD. The special operations unit in the water climbed into the RHIB, fired up the two-stroke outboard engine, and disappeared toward their mission into Kuwait.

During the Tethered Duck exercise the aircrew learned several valuable lessons, which would be incorporated into future operations. The moment the strap became fouled, the crew chief's keen procedural memory allowed him to immediately react with minimal thought. If he had been complacent, hesitated, or panicked, then there would have been a good chance of a major mishap, such as the downing of the aircraft and possibly killing some or all the crew onboard. The fact that the CSAR team was aware of the hazards and prepared to respond made the operation appear seamless. It's those types of lessons that can only be learned in the midst of chaos, and can then be brought back to develop better methods and procedures.

U.S. Navy Aviation Electrician's Mate Bill Rutledge (retired) flew 1,643 combat missions during the Vietnam conflict with Helicopter Attack Squadron (Light) 3 Seawolves (HA(L)-3). HA(L)-3 is the most decorated squadron in U.S. naval aviation history, and has served as the example and building block for CSAR tactics and procedures. Aviation rescue swimmer had not become an official U.S. Navy designation at that time, so aircrewmen like Rutledge were shaping the future of U.S. Navy helicopter operations. Rutledge was a senior aircrewman and door gunner, conducting back-to-back missions, as a quick reaction force for Navy SEALs and PBRs (patrol boat, riverine) in the Mekong Delta where VC (Viet Cong) and NVA (North Vietnamese army) funneled troops and supplies. This region where the Mekong Delta branches out to the South China Sea was a crucial area during the war. Six million Vietnamese, which was roughly 40 percent of the country, lived in this region making it both economically and strategically vital to the war-torn country.

During the Vietnam War, whoever controlled the waterways controlled the area, and initially this was the VC and NVA. In response, the United States developed Operation Game Warden to take control of the waters.

Most of Rutledge's missions were conducted as insertion and extractions of friendly forces, overland rescues, hot medevacs, aerial gunner support, body recoveries, and enemy body snatches. Primarily operating over land, Rutledge recalls a particular water search and rescue mission HA(L)-3 was diverted to. The rescue operation occurred after the U.S. Army lost ten helicopters during an assault on enemy forces, which was met with deadly resistance. Several military aircrews were either killed in action or wounded. The HA(L)-3 Seawolves' UH-1B helicopters joined up with an Army detachment to search for the downed aircrew.

The UH-1B Hueys took substantial enemy ground fire as they flew low over the area of where the battle occurred. One rescue bird took a direct hit to the main rotor and the helicopter aircraft commander immediately dropped the collective control and yelled, "AUTO! AUTO! AUTO!" The Huey violently slammed into the rice paddy 200 feet below, sinking and tilting into the swampy field. The rotors struck the water and ripped the helicopter around as the blades disintegrated, throwing wreckage for a hundred yards. The UH-1B lay still, upside down in a vulnerable enemy-flooded field.

One of the U.S. Army helicopters hurried in near the downed aircraft. The pilot jumped out with an M16 to provide cover fire while the aircrewman exited the helicopter to recover two victims. Once all were on board, the door gunner manned the mounted M60 and laid down fire as the pilot lifted and moved out of range. Due to extensive ground fire damage from enemy NVA, they communicated their status to the Seawolves aircraft and headed back to base. Rutledge's Huey then moved in as the primary SAR asset. The pilots brought the UH-1B into a low, slow hover over a rice paddy near the wreckage as Rutledge returned 7.62mm fire into the surrounding jungle. He then grabbed an M16 assault rifle that was hanging on the back of the HAC's seat and jumped into the neck-deep water. The muddy water filled his faded black leather combat boots and saturated his flight suit, but he didn't focus on his physical discomfort since he was taking on fire. The NVA enemy gunfire was like handfuls of gravel being thrown into the water surrounding him. With minimal regard for his own safety, he relentlessly drudged through the leech-infested water on his way to find survivors.

He approached the downed Huey, which was inverted in the deep murky water, making it impossible to determine if there were any survivors. Rutledge remained focused and kept his calm as he approached the unnerving metal wreckage. The downed helicopter's door gunner then suddenly surfaced from underwater, startling the searching aircrewman. Not realizing the gunner was already deceased, Rutledge began mouth-to-mouth resuscitation as bullets rained down on him, sparking off the Huey's fuselage. He then threw the lifeless victim over his left shoulder and sloshed his way through the deep water, unloading his M16 magazine along his path to the hovering Seawolves aircraft. He heard the WUMP! WUMP! sound and felt the forward pressure of multiple shots to the victim, draped over his shoulder. Then he felt the whiplashing impact of an AK-47 shot to his exposed back, knocking he and his victim face down into the muddy water. Rutledge, with the wind knocked out of him, treaded the spongy water to gain his breath and balance. Miraculously the lone bullet hit the "chicken plate" armor in his vest. Without spending a second to rejoice in his luck, he grabbed the victim and carried him toward the hovering Huey. It was difficult to distinguish between bullets hitting the water and the rotor wash pelting him in the face. The force pushed him back, but he leaned into it and slowly entered the calm underneath the rotor arc.

It wasn't until Rutledge reached the hovering helicopter that he realized the victim was dead. He quickly pivoted from rescue to recovery and strapped the door gunner's lifeless body to the UH-1B's starboard side skid. Completely exhausted, Rutledge pulled himself into the cabin and gave the pilot a thumbs up. He then manned the mounted M60, loaded a belt of 7.62mm rounds, and laid down suppressive fire as the pilots lifted the aircraft up and headed back to base.

Rutledge says he and his Seawolves aircrewmen were addicted to adrenaline. They all volunteered for the job and knew what they were getting into. They knew their job well and understood their life expectancy was low. However, they had so much wartime experience working in life-threatening and chaotic situations that they learned to channel their adrenaline to remain calm and focus on the mission at hand. They flew back-to-back missions, fluctuating from zero to a hundred as their fear and adrenaline were off the charts. The body must release those heightened endorphins at some point. When the daily operations were complete, it was very common to have uncontrollable body shakes and weep hysterically as they came down from their adrenaline high. The longer-term effects would surface years later after not working through the trauma, resulting in post-traumatic stress disorder (PTSD) and physical health issues. The selfless and heroic missions of the HA(L)-3 Seawolves are the epitome of leaning into adversity and the aviation rescue swimmer motto: So Others May Live.

* * *

> *"You may encounter many defeats, but you must not be defeated. In fact, it may be necessary to encounter the defeats, so you can know who you are, what you can rise from, how you can still come out of it."*
>
> —*Maya Angelou*

This quote describes leaning into adversity as all the heroes described in this book have done. Just as men and women must deal with the chaos and unpredictability of rescue situations, we all have this at times in our lives. Things can be uncertain and there might even be chaos. What we do in those adverse times predicts our mindset and ability to be resilient and come back stronger from life's setbacks and learn from them.

* * *

Senior Chief William "Bill" Rutledge was a true American hero. Bill received four Air Medals with V for combat mission, two Purple Hearts, two Bronze Stars with V for combat missions, the Navy/Marine Corps Medal, and two Distinguishing Flying Crosses. Sadly, one month after our discussion, Bill passed away. Our three-hour conversation will stay with me forever. It was an honor to hear his experiences and to have him listen with interest to my experiences as a "rotor brother." We spoke about faith, as Bill's Christianity was extremely important to him as it is for me. He was blown away by my Everest survival experience and how the hand of God saved me from certain death. He told me I needed to share my experience with everyone, just as he had been sharing his wartime experiences with others. He said that it was the only way to get our stories out, since when we die our stories die with us. I'm sharing one of Bill's stories but encourage you to search for him on the internet where he's written about other HA(L)-3 experiences, which have also been featured in other books and television segments. I am deeply saddened by the loss of Bill Rutledge but know that I will see him one day in heaven.

CSAR training is layered on to years of existing Aviation Rescue Swimmer training and continues to evolve as the U.S. military learns from their successes and failures. CSAR training and mission-sets are intense with constant unpredictable variables. No two missions are ever the same. They must plan for the unplanned and overcome impossibly dangerous obstacles. CSAR exemplifies the desire and willingness to lean into adversity, which makes it one of the most dangerous jobs in the world!

Chapter 9
Humility

During an unusually chilly 2013 summer night, Naval Air Station Whidbey Island's Search and Rescue (NASWI SAR) squadron was launched 80 miles east to assist with an evacuation of an injured forty-seven-year-old hiker in the Cascade Mountain Range. The incapacitated victim fractured his right leg near 4,800 feet from a fall during an attempt to summit Silver Peak Mountain. He was hiking with a group of friends when he slipped on the unstable talus field of boulders, snapping his right tibia. Fortunately, his hiking buddies were there and able to get out a 911 call. The popular climbing peak tops at 5,605 feet of elevation and is the tallest mountain on the southern range of Snoqualmie Pass. With 2,000 feet of prominence, it provides picturesque views of the surrounding Pacific Northwest mountains and lakes. Silver Peak has been personally one of my favorite training climbs for carrying increased weight in my pack for a good distance to prepare for other worldly adventures. Hiking with extra weight at a distance of over 8 miles helps increase strength and stamina to be able to handle larger bucket list mountains. In fact, Silver Peak was my last winter training climb in 2011 before I flew to Nepal to scale Mount Everest.

King County SAR coordinated the efforts between Seattle Mountain Rescue and NASWI SAR. Mountain Rescue was first on scene traveling by truck to reach the trailhead deep in the Snoqualmie Pass mountains. From there, the team set out on foot for 3 miles carrying their rescue equipment. Once onsite, they were able to stabilize the climber's injuries by placing his leg in a splint to prevent painful movement. Because of the distance, time of day, and complexity of the hike, they requested a helicopter evacuation. Helicopter Aircraft Commander Lieutenant Leah Tunnell, co-pilot Lieutenant Commander Fred Morrison, aviation rescue swimmers Melissa Dixon and Timothy Hawk, and Hospital Corpsman Brent McIntyre were on Ready Alert status when they got the call. They quickly geared up, launched in an MH-60S helicopter, and arrived at 10:15 p.m. The Mountain Rescue team was able to carefully guide the Whidbey SAR pilots to the injured hiker via continuous radio contact. The pilots dropped off McIntyre with some equipment lower on the mountain for him to meet up with the crew on foot. They then flew higher to prepare for a MEDEVAC extraction.

The aircrew spotted the headlamps on the pitch-black mountain and came into an approach for a hundred-foot hover. The pilots notified Crew Chief Dixon that they were having a hard time maintaining power due to the elevation and topology of the victim's location. They hovered in a mountain cirque, which is an amphitheater-like valley. A cirque can cause irregular wind conditions, making it difficult for a hovering helicopter to settle in one place to safely deploy Hawk and the rescue litter. The helicopter gauges were spiking in the red, which caused concern for the aircrew as the pilots fought against nature's resistance. Dixon, wearing NVGs, called out directions and altitude to the pilots so they could position the helicopter in a relatively safe area to insert the rescue swimmer. They planned to have Hawk do a nighttime rappel to a safe location a few hundred yards from the ground Mountain Rescue team, McIntyre, and the injured climber. Hawk would need to carry the rescue litter and medical supplies with him on the rappel rope. Once safely on the ground, he would hunker down long enough for the helicopter to reposition. Dixon and the pilots would essentially circle the area or find a place to land to conserve fuel and reduce the amount of debris kicked up by the rotor wash.

Hawk attached his locking carabiner to the rope and rappelled into the black abyss with the heavy rescue litter attached. Rappel rope naturally becomes twisted as it rapidly feeds through the rappel friction device such as a figure-8 or ATC belay device. This requires untwisting by the rappeler, as it can knot up and be difficult, if not impossible, to descend. This was the case as Hawk reached the halfway point, suspended 50 feet below the violent main rotors of the hovering MH-60S. He and Dixon were in constant communication using the airborne wireless intercommunication system (AWICS) as Hawk attempted to work out the problem. Dixon informed the pilots as they struggled with the aircraft controls to maintain a safe hover and not swing the helpless aircrewman into the tops of the surrounding coniferous trees.

Hawk was trying his best to quickly untwist the rope and descend to assist the injured climber. Over AWICS, he suggested cutting the rope so he could drop to the mountainous terrain below. Dixon contemplated his request but was conflicted about whether to comply since it was difficult to gauge how far he would drop. The dangling rescue swimmer thought he was much closer to the ground than he was since he could sort of make out a downed tree just below him. But his judgment was faulty, as the dark forest and shadows mixed with the forceful winds of the rotors made the earth appear closer than it was.

As the crew chief, Dixon felt a wave of panic flash over her as she assessed the situation to make a quick decision, a decision that could turn catastrophic for her friend, Hawk. In life sometimes we live or die by the decisions we make with the limited information we have at the time. It's the best choice based on the information, but disastrous when things don't work out as planned. Others who weren't there will always question you as if they know better, but nobody will ever question the decision more than yourself. If the outcome is tragic, then you will carry immense guilt and be haunted by that split-second choice the rest of your life. You would do anything to have a chance for a redo. But you can't change those decisions or the past. You can only work on the present and future. Most importantly you will need to find forgiveness for yourself to move forward.

The pilots were struggling to keep stable helicopter power. Hawk was stranded 50 feet below the helicopter and suggesting cutting the line. Dixon assessed the situation and calculated the risks involved in their actions; then she ordered Hawk *not* to cut the rope! Instead, she had him release the litter to get a better indication of his distance from the ground. Hawk compliantly followed his crew chief's directions and detached the rescue litter. An eerie chill shot through his body as he watched it fall and smash the log below. However, it didn't stop there. It continued falling well beyond the log, over a steep embankment and out of sight into the dark forest. Dixon's decision to *not* cut the rope was the correct decision. Hawk agreed.

The twenty pounds of weight released from the detached litter provided enough relief on the rope for Hawk to untangle the twisted knots. Once the rope was free of obstruction, he rappelled the remaining 50 feet. As he detached the rope from his rappelling device and alerted the crew chief that he was clear, he looked down and noticed the mangled litter. He breathed a sigh of relief and then moved to a secure location to allow the helicopter to depart. With boots on the ground, he hiked down to the location of the fallen victim, McIntyre, and the Mountain Rescue team.

Along the descent, the Navy rescue team salvaged the rescue litter and were happy to see that it just sustained some dents but was still fully functional. They assembled it at the triage station while the Mountain Rescue team provided a complete synopsis of the climber's condition, which they relayed to the MH-60S aircrew above. McIntyre and Hawk secured the exhausted climber in the litter and called for a helicopter extraction. The HAC flew back into a position closer above them and Dixon hoisted McIntyre and the victim up first, and then Hawk with the remaining medical supplies with no further incidents.

There are so many variables in a rescue that need to be constantly considered to impact the safety of the personnel and mission. Unexpected obstacles will almost always occur during a SAR mission. And they tend to surface during life-threatening moments, which can turn extremely dangerous in a matter of seconds. As crew chief, Dixon was faced with a split-second decision based on her limited data, critical timing, and environmental impediments. Dixon, a former surface rescue swimmer turned aviation rescue swimmer and top-ranking Crossfit Games athlete, relied on her training, experience, gut instinct, and ability to remain calm in chaotic situations. She rejected the initial suggestion of the suspended rescue swimmer and quickly provided a possible alternative solution. Cutting the line would have resulted in a substantial night plummet, adding to the victim count for that mission. Hawk had the noble attitude of risking his life to save another, but he didn't have a clear picture from his vantage point. This is very common when we're too close to the problem. When we find ourselves in that place, it's important to pause and take a mental step back. This gives us a wider view to deal with the situation from a better perspective. In this example, Dixon had the wider view and fortunately made a life-saving decision.

* * *

Perspective is defined in the Oxford Dictionary *as "a particular attitude toward or way of regarding something; a point of view."*

Perspectives are person-specific; usually we only see our own point of view. When we take on other's perspectives to try to understand where they are coming from, it increases our empathy, concern, and emotional intelligence. It also requires a bit of humility. Sometimes we must remove our own emotion to do so or even take a step back to come from a more detached position or broader view such as in the above rescue mission. In everyday life we can broaden our perspective by surrounding ourselves with people who are different than us. We can pay attention to others and become involved in their experiences. If this part alone is difficult, we can start by watching TV or reading about people or experiences that are different than what we are familiar with to increase our understanding of others. Ultimately, everyone's goal should be to learn to love one another in our diverse world.

* * *

The Aviation Rescue Swimmer Course (ARSC) is designed to provide U.S. Navy and U.S. Marine Corps personnel with the knowledge and skills necessary to initially qualify as search and rescue swimmers. Training includes water entry, advanced first aid, CPR, parachute disentanglement, and waterborne life-saving procedures, under all possible environmental conditions. This also includes successfully performing recovery procedures via hoist or over land to rescue stranded or injured personnel. As the Navy's primary SAR asset, the aviation rescue swimmers will be launched on overland rescues in addition to overwater rescues. The land rescue portion of training used to occur in the fleet, months later in the aviation rescue swimmer's career. However, due to the increasing overland SAR mission-sets conducted in the fleet, it made sense to provide the candidates the building blocks during Aviation Rescue Swimmer School (ARSS). The overland training includes classroom lectures to learn the different carries they will use in the field, including fireman's carry, chair carry, drag, and litter carry. They then practice the carries with the other candidates playing the victims, plus carry 75-pound dumbbells 40 feet to simulate carrying a person on a litter with gear on. More in-depth, onsite training will occur in the helicopter when the students successfully complete the almost two-year training program and are assigned to a squadron.

The squadrons are based all over the world but all carry the same standard SAR mission task. Depending on their designation they will focus on more specific mission-sets, but the core function of Navy helicopters stem from being a SAR asset, both over water and over land. Each squadron must maintain operational readiness by continually keeping their qualifications current. The aircrew trains in the squadron helicopter hangars to become proficient in the land rescue gear and procedures. They then take to the air for simulated rescues over land. They may use a training dummy for the victim or drop personnel in a location to create a more realistic scenario. Since many land rescues are in environments not conducive to landing a helicopter, they practice deploying down the hoist during a hover and extracting the victim via a rescue litter. This involves meticulous coordination and communication, with a very small margin of error. The aircrew will conduct the training and then regroup back at base to debrief and document lessons learned. Navy SAR teams don't accept anything less than perfection as they qualify to save lives around the world in domestic and hostile environments.

Additionally, the U.S. Coast Guard created an advanced Rescue Swimmer course in 1995, located in Astoria, Oregon. Only eight classes are conducted each year in the fall and winter, with instructors including pilots, aircrew, and aviation rescue swimmers from the U.S. Coast Guard, U.S. Navy, U.S. Air Force Pararescue, and Royal Canadian Air Force. The Oregon coast was chosen due to the consistent high seas, cold water, cliffs, and caves required for the intense five-day program. It's a humbling experience to be at the mercy of Mother Nature but trained to navigate within her unforgiving boundaries to save lives.

Each day consists of classroom instruction and briefings, followed by onsite practical application. The Coast Guard instructors first perform the training evolution to demonstrate the techniques in action, then the students perform the same tasks. After each day's training, they regroup to discuss what went right and wrong to build on their lessons. The aviation rescue swimmers have the opportunity to swim and perform rescues in cold water and high seas, wearing dry suits and full rescue gear. Dry suits are essential due to the extreme cold temperature of the Pacific Ocean, which can cause a person to become hypothermic within seconds. The pilots and aircrew learn techniques to safely deploy and extract the swimmers and survivors in harsh ocean conditions. They learn to time the waves or sets of waves to ensure they minimize natural dangers that can quickly overcome an operation.

U.S. Navy aviation rescue swimmer Joe "Q-tip" Sutherland attended the Coast Guard Advanced SAR school during the fall of 2000. Originally from the Pacific Northwest, he had plenty of experience swimming in the Pacific Ocean. He and I grew up as best friends in a small town in southern Oregon named Rogue River. We spent our youth exploring the Siskiyou Mountain range and swimming in the surrounding rivers and at Harris Beach in Brookings, Oregon. I enlisted in the Navy shortly after graduating in 1992 while Joe attended community college. After graduating ARSS, I visited him in Rogue River and encouraged him to join the military. A year later he was going through the arduous aviation rescue swimmer pipeline.

While I was stationed in San Diego, Sutherland protected the East Coast flying in SH-60B helicopters while stationed in Jacksonville, Florida. He operated in the Atlantic Ocean for five years, quite a contrast from the frigid temps of the Pacific. And although the East Coast has its share of deadly predators like great white sharks, they are not as intimidating as the Pacific Northwest's large population. The Oregon coast is a popular area for great whites due to the high number of seals and sea lions, which they regularly feed on. During training and prior to an open ocean rescue litter evolution, the Coast Guard instructors happily briefed the students on the potential wildlife they would be working near that day. It's a good reminder that we are borrowing time in their environment since they ultimately live there.

An aviation rescue swimmer's job is to get in and out as efficiently and effectively as possibly. That's why the helicopter is the vehicle of choice with its speed, maneuverability, and flexibility to perform rescues anywhere at any time. However, the high-pitched engine noise, thumping rotor blade sound, and forceful rotor wash has been known to attract sharks. If that doesn't draw the great whites in, then a rescue swimmer in the water, wearing a black wetsuit or dry suit and fins, looks awfully close to a shark's favorite meal from the dark depths below.

Sutherland was the lead SAR training petty officer for his East Coast–based helicopter training squadron. With countless hours teaching, training, and placing survivors in the rescue litter, he was pumped to perform the task a mile directly west of Oregon's northern rocky shores. He and a fellow Navy aviation rescue swimmer, plus a Navy pilot, briefed with the Coast Guard rescue swimmer and pilot instructors. They went through the details of what to expect and how to operate in the cold, high Pacific seas. The crews donned their dry suits and carried their rescue equipment bags to the Coast Guard HH-60J Jayhawk helicopter, conducted the pre-flight checks, and prepared the cabin for a rescue litter hoist evolution. The aircrew were nervously excited to reach their destination and get into the water for some realistic open ocean training.

The flight was shorter than expected as the Jayhawk arrived on scene, slowed down, and dropped in altitude. The Coast Guard flight mechanic hoist operator opened the cabin door and tossed out a MK-58 MOD 1 smoke. The pilots pulled the helicopter into a SAR pattern, flying an extended circle back to the smoke, and dropped to 10 feet / 10 knots into the wind. Both rescue swimmers jumped from the moving helicopter and gave the OK hand signal. The Jayhawk then lifted up to a 70-foot hover and moved back and to the left. The two advanced SAR students then alternated roles as the survivor and rescuer. Sutherland began the training as the first rescue swimmer and immediately conducted his checks on the simulated survivor. Noticing a possible broken back he signaled to the hoist operator the need to deploy a rescue litter by raising one arm with open palm and fingers extended over the rescue swimmer's head and using the other arm to touch the raised arm at the elbow. From the hovering helicopter it looks like an L for litter.

The sunlight glimmered down through the murky water as Sutherland carefully kept the survivor afloat, while the hoist operator configured the rescue litter and trail line. He then lowered the weighted trail line to Sutherland who grabbed hold and kicked backward to ensure it was taut as the flight mechanic lowered the litter connected to the rescue hoist. The trail line keeps the litter stable and reduces spinning while being hoisted up. Once in the water he disconnected the litter from the rescue hook and signaled to the crew chief that it was clear. Sutherland positioned the litter and survivor clear of the rotor wash, by working with the survivor's back toward the hurricane force winds. This allows the survivor's airway to remain clear while the rescue swimmer, wearing a mask and snorkel, secures him into the floating rescue device.

Sutherland worked proficiently to place the survivor in the litter and connect the proper restraint straps, beginning with the first one under the survivor's arms and over his chest to secure him in the buoyant litter. The litter has flotation to ensure the survivor's head remains above water when properly strapped in. If not properly strapped in, then the survivor risks flipping over and being pinned face down in the water or slipping out during hoisting up to the helicopter. Neither are good options. The final black restraining strap locks the survivor's arms down to prevent any slippage. Sutherland made sure the sling cables were free and clear, and gave a thumbs up to the flight mechanic. As he hoisted the litter up from the ocean, the rescue swimmer paid out the tight trail line to keep the litter stable as the survivor went for a 70-foot ride to the hovering Jayhawk helicopter.

He watched as the flight mechanic brought the survivor and litter into the helicopter's cabin and then pulled up the trail line. Sutherland treaded water at the two o'clock position outside of the rotor wash and pondered about the creatures lurking below. The HH-60J then departed to fly the SAR pattern to deploy the rescue swimmer to now rescue Sutherland, who would assume the survivor role. As the helicopter flew away, he bobbed in the chilly water while playing out the different sea life scenarios in his mind. He recalls wondering whether it would be better to see the great white coming or to just get attacked without warning. Fortunately, Sutherland didn't have to deal with either option, as they didn't encounter any sharks. The fear of the unknown is real, and the unpredictable ocean easily humbles the most confident and capable rescue swimmers. Don't let them tell you anything different. Focusing on the task at hand helps to temporarily compartmentalize those fears. It doesn't help or make sense to focus on the fear, since there's not much you can do about a shark eating you anyway.

* * *

Fear is an important response to physical and emotional danger. There are many times when we need fear because it keeps us safe, and we wouldn't protect ourselves if we didn't feel and respond to fear. However, there are many times when a perceived threat is not reality, and our thoughts can overtake our life. When fear is crippling us with anxiety or panic, it's time to seek help. First, a medical doctor to make sure there's not a medical reason for the increased fear, as there are numerous conditions that increase cortisol in the body, which contributes to feelings of panic or anxiety. Second, find a therapist that can help with cognitive behavior therapy (CBT) to retrain patterns of negative thinking and help challenge and reframe harmful beliefs.

A simple question to ask yourself is, "Is this true?" Often with fear we start to believe lies. Recognizing thoughts and not accepting them as fact can be a powerful step in overcoming irrational fears. You alone have control over allowing the thought to stay or "letting it float away" as I like to call it when using the mindfulness technique of visualization. As an example, you can watch your unwanted thought be put inside a red balloon and watch it float away. Visualization can also be used with reoccurring negative or intrusive thoughts to distract yourself and practice mindfulness. My place of visualization is a beach in Hawaii, and Brian's is probably on top of a mountain somewhere. It just needs to be a place of peace for you. You can even print a picture and place it where it might be needed, but mostly it's a safe place in your mind that you can go to whenever you need to get to a state of calm.

* * *

Vertical surface or cliff rescue training is conducted on the rugged, windswept cliffs of the Pacific Northwest. The Oregon coast has been working on its picturesque formations for millions of years. Old lava flows cooled into solid basalt and millions of years of erosion have created the magical formations that Lewis and Clark explored during their famous expedition in the early 1800s. Several movies have been filmed with the natural majestic backdrop, such as *The Goonies*, *Twilight*, and *Point Break*. With the steep cliffs, high winds, and crashing waves, it makes for a perfect location for the Coast Guard's Advanced SAR training.

Sutherland and his aircrew briefed at the Astoria base and then flew in a HH-65 Dolphin helicopter to their training area located on a 300-foot cliff weathered from the relentless winds. Below, the ocean raged and crashed against the sea stacks and rocky shore. The Coast Guard instructor first lowered down to the cliff to set a rescue dummy, which is a 180-pound replica of a human. The pilots learn to hold a tight hover in the high winds over the cliff, using their instruments to determine the wind patterns. The flight mechanic lowers the rescue swimmer to the stranded victims on the cliff to either be hoisted in a rescue strop or litter, depending on the severity of their injuries.

Sutherland connected his harness-lifting V-ring to the rescue hoist and was lowered down to the side of the cliff. The goal is to have the rescue swimmer remain attached to the rescue hook while the flight mechanic guides him/her vertically and horizontally. Sutherland used a series of hand signals to direct the hoist operator where to move the helicopter to get him closer to the survivor. The flight mechanic continually communicates to the pilots and watches for any hazards that could impact the main and tail rotors. The rescue swimmer never detaches from the cable, which feels unnatural, but instead is guided like a puppet from the loud, hovering aircraft.

Sutherland recalls the helpless feeling of being tethered to the cable when a large wind gust shifted the helicopter and ripped him 15 feet away from the cliff. He experienced that sinking feeling you get when going on a roller coaster. And I know he doesn't like that feeling since he threw up on every roller coaster we went on as kids. The pilots and flight mechanic do everything in their power to keep the helicopter and rescue swimmer steady, but Mother Nature doesn't always comply. They slowly hovered back to the cliffside with Sutherland hanging on the wire, where he gained purchase of the rocky terrain and moved closer to the rescue dummy. He pushed aside the deafening engine and rotor noise, nerve-wracking heights, the whirlwind of debris from the powerful rotor wash, and everything outside of his control. Instead, he zoned in on the simulated survivor needing his help. When you serve others and place their needs above your own, it's amazing what you can accomplish. It just takes gaining new perspective and purpose by refocusing your thoughts and actions.

Attached to the hoist and in a steady hover, Sutherland climbed up to where the rescue dummy was laying and established communication. It was a one-sided conversation. He then conducted a thorough medical check, searching for life-threatening injuries. With no spinal injuries present, the rescue swimmer strapped the survivor securely into the rescue strop. He leaned back to make eye contact with the flight mechanic and gave a thumbs up for them to both be hoisted up to the hovering Dolphin helicopter. During vertical surface rescues, the rescue swimmers are not only operating in a dangerously unstable and unpredictable environment, but they must also work efficiently to minimize the time on scene.

During another day of training, the students spend time learning the dynamics of performing rescues in sea caves and from wet, coral-coated rocks. Helicopters aren't used for this training, but the rescue swimmers go inside an actual sea cave being hammered by the ocean waves. They learn how the water moves in and out of the cave and to use its power to safely enter the cave, navigate inside, and egress back to the open ocean. The Coast Guard Advanced SAR instructors conduct exercises rescuing victims who are stranded on sea stacks or rocks. The students learn to use the crashing waves to their advantage to ride and propel with the rough current to reach trapped survivors. They then time the waves to safely extract the survivors out beyond the rocky hazards where they either find a safe path to shore or get hoisted up into the rescue helicopter.

The week's final training evolution is to practice survival techniques after a simulated helicopter crash. This scenario is as realistic as it gets with the cold, high winds, and waves of the Pacific Ocean. The students jump from the HH-60J Jayhawk and inflate a life raft. They then learn to use the raft and demonstrate survival equipment usage. The pilots and flight mechanics involved in the training also get hoisted to the helicopter during this exercise. This provides them with the experience of what the rescue swimmers do for a living.

The Coast Guard Advanced Rescue Training takes the standard water and land SAR procedures to the next level with very realistic conditions and scenarios. Everyone that I've spoken to who's had the opportunity to attend the training has given it great reviews. They felt it enhanced their knowledge and confidence as an aviation rescue swimmer, which they were immediately able to model in the fleet. Since only a few U.S. Navy aviation rescue swimmers can attend each year, they bring their experiences back to their respective squadrons or bases to train others.

* * *

Mount Baker, the third highest mountain in Washington State, looms at 10,781 feet in the North Cascades Range. Mount Rainier (14,411 feet) is the tallest mountain in Washington with Mount Adams (12,281 feet) coming in second. Mount Baker is the second most thermally active volcano in the Cascade Range after Mount St. Helens, which had a devastating eruption in 1980. Baker has the heaviest glacial coverage of the Cascade volcanoes, and its location just east of the Puget Sound makes it one of the snowiest places on Earth. In fact, it set the world record for snowfall in a single season in 1999 when it accumulated 1,140 inches (95 feet) of snow. Its immaculate presence serves as a backdrop to surrounding cities and can be seen from as far south as Tacoma and as far north as Vancouver, British Columbia. Two former U.S. Navy ammunition ships, USS *Mount Baker* AE-4 and USS *Mount Baker* AE-34, later redesignated USNS *Mount Baker* T-AE-34, were named after the mountain.

Mount Baker is a popular destination for skiers, snowboarders, and mountaineers. I have personally climbed it several times on different routes to train for other high-altitude expeditions or to lead a team of individuals looking to increase their skills and confidence prior to attempting larger peaks like Mount Rainier or Denali. The steep terrain and unpredictable glaciers, crevasses, and icefields make it an ideal location for an epic mountaineering experience. However, with those conditions comes risk, which can also be part of the appeal. I have conducted a few different rescues on the mountain, where I happened to be at the right place at the right time. I've had the opportunity to assist less experienced individuals down through the gauntlet of snow bridges and split open glaciers as well as pull out experienced climbers from accidental crevasse falls.

During the summer of 2000, a three-day heat wave made the snow above 8,000 feet into perfect slab avalanche conditions. This is where a bonded layer of snow or slab lays on top of a weaker layer. The slab, typically at a slope angle of at least 30 degrees, can easily be triggered by a low additional load such as falling rock or ice, weather, or humans. Summer is the most popular time to climb the larger Cascade peaks due to the more predictable weather and settled in snow conditions and routes. However, even in ideal conditions, objective hazards remain. Mountaineers should wear proper safety equipment, like a harness, rope, helmet, ice axe, crampons, and avalanche beacon. They should also possess proper skills, leadership, and practice avalanche avoidance and response.

On June 25, 2000, experienced climbers forty-year-old David Pougatch and thirty-five-year-old Andre Boulanov set out to scale Mount Baker via the more technical and less climbed North Ridge route. Pougatch and Boulanov were members of the first Internet Russian Alpine Club and were residents of Washington State. The roped-up climbers were ascending the prominent ice wall of 60–85 degrees about 1,500 feet below the summit. Due to the steep pitch and risk of falling, the climbers used ice screws for protection, which they clipped their rope into for a running belay. Both carried two ice axe tools, and they drove the pick into the vertical ice as they kicked in the front points of their crampons to maintain solid purchase as they moved higher.

The details of the accident are limited, but at some point on their ascent the tethered climbers either slipped on the bullet-proof ice or broke loose an ice screw and fell 100 feet. Both climbers helplessly fell straight down to a hard snow shelf below and, after a forceful blow, began a treacherous slide down a 60-degree slope. Pougatch accelerated down the glacier wall and then disappeared into a deep crevasse. The Russian mountaineer died instantly on impact. Pougatch's body, lifelessly wedged in the crevasse, halted Boulanov's uncontrollable descent. During the frenzied fall Boulanov broke both arms, legs, several ribs, and also suffered from a punctured lung and head injury. With one deceased, the remaining climber hung barely conscious in critical condition.

A four-person climbing team celebrating their own success on the summit witnessed the shocking incident. Horrified by what they had just observed, they called 911 from a cellphone. The team cut their summit time short and began planning their descent to help the fallen mountaineers. They needed to find an alternative route that was safe enough for each of their individual skillsets as well as an area to avoid kicking down debris on the injured victims. They were exhausted from their climb, but knew they had to be diligent on their descent as most accidents occur on the way down. It would take quite some time for them to reach them, which would leave Boulanov alone for hours in his critical state.

The Whidbey SAR aircrew, consisting of Crew Chief (ADC) Frank Leets, Medical Technician Senior Chief Petty Officer Bryce Schuldt, Aviation Warfare Systems Operator First Class Petty Officer (AW1) Marty Crews, and pilots Lieutenant Commander Scott Parish and Lieutenant Chris Cote, were stationed at Naval Air Station Whidbey Island. While scheduled as the Alert Crew they received a distress call from the accident on Mount Baker. The initial report was vague and all they knew was two climbers attempting to climb the 10,000-foot volcano had fallen into a deep crevasse near 9,000 feet. They were informed that one of the climbers was suspected to be deceased and another was in critical condition.

The aircrew quickly briefed with their limited intel while they prepped their equipment for a rescue and possible body recovery. The U.S. Navy UH-3H Sea King helicopter, callsign Firewood Six, spun up the main rotors and launched from NASWI. Enroute to the massive Cascade mountain, they landed in Bellingham to pick up Delvin Crabtree, a local Bellingham Mountain Rescue Council spotter. Crabtree had been previously briefed on the incident and provided additional details to the Navy aircrew of the area and objective hazards to avoid. With years of experience on the mountain, he had firsthand knowledge of the dangers around the area the climbers fell. They would have to avoid a massive ice wall due to the altitude and steepness, plus be cautious that the noise and power of the rotors could kick off life-threatening avalanches.

The thirty-year-old Sikorsky helicopter nearing its end of life flew east toward Baker and was about to be tested beyond its limits. They approached the blinding white scene of the afternoon sun banking off the glacial ice and came into a slow approach. Due to the thin air of the altitude and high vortex winds whipping down from the northeast, Lieutenant Commander Parish and Lieutenant Cote had to quickly change their tactical plan of engagement. The old, heavy aircraft struggled to maintain a stable hover at 9,000 feet. The main rotors ferociously spun, but due to the weight and thin air the engines couldn't provide the necessary torque to safely perform at altitude. The pilots decided to fly a lap to a safe area to jettison some fuel and then descended to a safer ridge a thousand feet below the stranded climbers. Once in a steady hover, the crew chief hoisted down the three rescue personnel to the dangerously vertical Roosevelt Glacier. The pilots of Firewood Six then set its course back to Bellingham International Airport to refuel.

Schuldt, Crews, and Crabtree hunkered down and protected their rescue equipment from the rotor wash as the helicopter dropped down into the valley and disappeared to the north. Looking up at the blinding white ascent, they saw they were facing an extremely steep 45-to-60-degree slope that climbed up 1,000 feet of treacherous glacial ice, which was littered with deep crevasses. Schuldt lugged the medical supplies while Crews and Crabtree carried a rescue litter. Lacking critical mountaineering gear protection such as crampons and ice axes, they kicked in steps and slowly inched a path up the snowy mountain. Evidence of recent avalanches framed in their narrow passage toward the critically injured Boulanov. The afternoon temps beat down on the mountain slope, making each step less stable than the last. They had to ensure they had secure purchase of their boots; otherwise, they could easily slip and become victims themselves. With the blazing hot summer heat, their frustrating steps would pack down deep in the warm snow and then suddenly give way. Several times they would sink to their thighs and then uncontrollably slide back down the mountain. During normal conditions the climb would take a half-hour, but on that Sunday, it took over four exasperating hours.

The rescue team eventually reached the top of a ridge at 8,400 feet just to be faced with an impassably deep and wide crevasse edged by 30-foot ice cliffs. With the clock ticking, they decided to rig a solution by setting a snow anchor on their side of the crevasse and having Schuldt free climb while connected to the climbing rope, through the glacial obstacle to then set an anchor on the far side. Snow anchors are built during mountaineering or rescues to provide added protection when traveling on steep slopes, unstable environments, or aiding in crevasse rescue. They are typically set with a device such as a metal picket, ice axe, or ice screw to then anchor the rope via a locking carabiner. Since the rescuers lacked essential mountaineering rescue devices, they had to improvise and use whatever resources they had available to secure the rope across the crevasse. Utilizing the solid ice of the glacier and leverage of angles and weight can be an efficient and effective method of building a temporary anchor. Crews belayed Schuldt across the void with a safety line attached in case of a fall. Schuldt then set the far side anchor protection and scrambled up the remaining 600 feet to meet the critically injured Boulanov. Crews and Crabtree traversed the crevasse with the rescue litter and met up shortly with Schuldt, Boulanov, and the other climbers who had descended from the summit to help.

The Navy rescue team assessed the situation, confirming that Pougatch didn't survive, and conducted full medical checks on Boulanov. They carefully stabilized his neck with a cervical collar and secured him into the rescue litter. It took all three men to lower him down the steep slope, due to the weight, steepness, and unstable terrain. After an hour of intense work descending toward the large crevasse, the injured climber's breathing and heart suddenly stopped. Schuldt recalls saying, "You're not dying on me now, not after all this!" He began CPR on Boulanov as Crews radioed Firewood Six for immediate extraction. Crews recalls telling the helicopter crew chief, "The guy's coded! It's now or never!"

Parish and Cote flew a pattern below the extraction point, expecting to pick them up at 8,000 feet. As soon as they heard the radio call about a higher-altitude rescue, they diverted to jettison more fuel to reduce the helicopter's weight. On the final approach to the side of Mount Baker, Crew Chief Leets called in the distance and elevation as the pilots slowly maneuvered the aircraft closer to the rescue team below. Schuldt, Crews, and Crabtree held low and tight in the exposed environment as the rotor wash pounded down, threatening to knock them off balance. The UH-3H inched up the mountain with the 30-foot rotor blades visibly struggling in the thin air. At times, the blades came within 3 feet of the glacier. A rotor strike at that altitude and in those conditions would have resulted in a most certain game over for the aircrew. The helicopter could barely hold a hover as the crew chief hoisted up Schuldt and Boulanov, strapped in the litter. Crews and Crabtree remained on the side of the mountain as Firewood Six departed back to Bellingham.

In the helicopter cabin, Leets and Schuldt secured the litter and alternated performing CPR. Schuldt used the defibrillator three times to shock Boulanov's lifeless body back to life. The flight to Bellingham hospital only took fifteen minutes, but they landed with only five minutes of fuel remaining. On the hospital's helicopter pad, medical technicians quickly transferred the critical victim from the UH-3H to ICU. Schuldt and Leets briefed the hospital dispatchers on Boulanov's condition and what they had performed on him on the mountain and in flight.

Firewood Six refueled and then hustled back to Mount Baker to extract Crews and Crabtree. The two had descended lower to help reduce the risk of the extraction, plus to clean up their snow anchor and rescue gear. Both aircrew had been on the mountain for over eight hours with limited supplies or water. They suffered from extreme dehydration and Crews was experiencing snow blindness from the sun's glare banking off the ice, which essentially sunburned his cornea. Firewood Six approached into a slow hover and the crew chief hoisted them both to the cabin. As the helicopter departed back to Bellingham and then Whidbey Island, Crews and Crabtree collapsed with exhaustion. After a much-needed night of rest, they returned the following day to recover David Pougatch's body from the crevasse.

The Whidbey SAR aircrew wasn't trained for high-altitude glacier travel, mountaineering, or crevasse rescue. Additionally, the UH-3H helicopter wasn't equipped for a rescue at that altitude or thin air. Despite the less-than-ideal conditions, the team ignored their own safety to conduct a miraculous rescue and recovery. Unfortunately, six days later Boulanov passed away. The pilots, aircrew, and Mountain Rescue member were highly trained and experienced to work through extreme conditions and scenarios. However, that day they operated far beyond their limits and adapted to each obstacle they encountered. The entire team showed their humility and selflessness by stepping up to overcome dangerous challenges. Both Schuldt and Crews were later awarded the Navy and Marine Corp Medals with noted extraordinary heroism. Although both climbing victims passed away, Crews said it was all worth the rescue efforts since Boulanov was able to at least talk to his family before dying, and they were able to provide closure to Pougatch's family by returning his body home.

Although the victims didn't survive, the rescue was deemed one of the best, most difficult they've ever seen. "On a scale of one to 10, that was a 10-plus," said Chief Ron Peterson, a thirty-year mountain rescue veteran who oversees SAR rescues for the Whatcom County Sheriff's Department. "It was the worst time of day for us, the worst technical conditions, and very critical conditions. The whole crew did a marvelous job."

* * *

Those who practice humility are more empathetic and compassionate because it requires you to consider others and not deem yourself better. To authentically serve others requires humility to see all people as deserving and equal. This also helps us continue to grow through self-awareness because humility allows us to see and examine who we truly are.

* * *

Land rescue training is a critical element in the Aviation Rescue Swimmer School curriculum. Although the rescue swimmers will mainly be deployed on Navy warships, there will always be rescue needs beyond the water. Proper training and maintaining qualifications ensure the aircrew always remain in an operationally ready status. The Coast Guard Advanced Search and Rescue training humbles the candidates by subjecting them to advanced tactics demonstrated in harsh environmental conditions. The location they train in is intensified by the powerful waves and freezing temperatures of the Pacific Ocean. Aviation Rescue Swimmer School is thorough, covering a lot of pertinent scenarios, but the Coast Guard Advanced training is next level in providing more intense and unpredictable terrain to hone the rescue swimmer's skills. Rescues rarely go as planned, which is why continual training in different environments builds confidence in the aircrew to respond and perform no matter what the call.

Chapter 10

Focus

On November 14, 2002, Lieutenant Junior Grade Duane Whitmer and aviation rescue swimmer AW3 Richard Pentony were as green as they come when they briefed for their first training flight together in Helicopter Anti-Submarine Squadron 10 Warhawks (HS-10) Fleet Replacement Squadron located at Naval Air Station North Island (NASNI) in Southern California. Whitmer was flying his first overwater stage event, while Pentony was flying his second training flight, which is an introduction to SAR events. Fortunately, they were both flying with a seasoned aircrew, who previously served together at HS-14. Commander Dan Kletter was the COMHSWINGPAC operations officer flying as the instructor pilot and the instructor aircrewman was AW2 Evin McDonel. The HS-10 aircrew was flying Calumet 621, a fully equipped ASW and SAR SH-60F.

Weather wasn't ideal that cool, foggy morning off the coast of San Diego. Calumet 621 was assigned to a location called the Echoes, where they were able to conduct their training in visual flight rules (VFR) conditions. The Echoes are one of the training quadrants located miles west off the Pacific coast between NASNI and Imperial Beach. VFR is an aviation set of regulations under which a pilot can operate the aircraft in conditions clear enough for them to see where the aircraft is going. Instrument Flight Rules (IFR) is the opposite, where the pilot must fly only using instruments due to having no visual references. Less than an hour after takeoff, Calumet 621 completed two practice day SARs and two practice night SARs when a distress radio call came in from "Beaver"

Fleet Area Control and Surveillance Facility (FACSFAC) San Diego. Beaver urgently requested for any SAR-capable helicopters in the area to respond, because a U.S. Marine Corps F/A-18 pilot had just ejected 52 nautical miles south of their position. Calumet 621 responded.

Beaver confirmed to Calumet 621 that they were the closest SAR asset and provided coordinates for them to divert toward. The F/A-18 wingman, Shooter 01, of the downed aircraft and a P-3 Orion on scene coordinator (OSC) flew overhead of the USMC pilot in the water, remaining on radio contact. The OSC reported that the pilot was in a raft and appeared to be in stable condition. The Calumet 621 pilots inputted the survivor's coordinates and flew a direct path at 150 knots as Pentony donned his SAR swimmer gear and McDonel rigged the cabin for rescue. Pentony recalls the adrenaline surge charging through his body as he searched his SAR bag for his booties. His mind raced as the deafening, high-frequency squeal of the main rotors spun overhead. It was surreal, and he kept wondering if this was really happening on his second actual helicopter flight. The crew chief calmly looked at him while holding his booties and yelled, "Are you looking for these?" At that moment, Pentony paused and smiled, reaching for his booties. He immediately felt a wave of calm and went directly into Aviation Rescue Swimmer School training mode. He pulled on his booties and fins, placed his mask with snorkel on his head, and scooted toward the open cabin door. The young rescue swimmer was locked in and ready to save lives.

Calumet 621 approached within 1 nautical mile of the downed aircraft, but due to the foggy conditions didn't have a visual contact of the pilot in the water. Kletter radioed the P-3 OSC, requesting to have the pilot ignite a marine smoke, which they relayed to the floating survivor. The crew chief immediately spotted the orange smoke pouring from the handheld MK-124 MOD 0 and trailing bright green sea dye marker staining the ocean's surface. The pilots turned the helicopter to the starboard and communicated their approach. They then dropped to an altitude of 10 feet at 10 knots as Pentony sat in the doorway, legs and fins hanging out, ready to deploy. Kletter called out, "JUMP! JUMP! JUMP!" The crew chief gave three consecutive taps to Pentony's right shoulder and the rescue swimmer exited the helicopter to the 60-degree ocean water below. Pentony cleared his mask underwater and surfaced with an OK hand signal. He then turned and swam toward the survivor as the SH-60F moved left and back while increasing their hover to 70 feet.

The rescue swimmer approached the shivering F/A-18 pilot, communicating his intentions as he assessed the situation. The pilot's seat pan and parachute had automatically detached by firing their saltwater activation mechanisms and were clear of the survivor. A layer of jet fuel floated on the surface of the water and the tail of the fighter jet was visibly bobbing in the low sea-state 50 yards away. Pentony calmly communicated with the pilot as he conducted physical and visual checks to ensure the pilot wasn't suffering from a back or neck injury. He then removed the pilot from the raft by pulling the pilot's harness from behind and kicking backward as the flotation device slid out from underneath him. After ensuring the pilot's combined torso and harness vest's flotation was functional, he unsheathed his SAR knife to puncture and sink the raft. He recalls stabbing the raft multiple times with the dull blade before finally penetrating the military-grade rubber. The rescue swimmer then tried a few times to stow the knife into its scabbard on his harness, before just tossing it into the ocean. There have been several cases where rescue swimmers have wounded themselves in an attempt to save a $50 knife. Sometimes it's better to prioritize completing the mission over attracting sharks to the rescue.

Pentony gave a thumbs up to the SH-60F crew chief and they immediately moved in for extraction. McDonel lowered the rescue hook and within a minute Pentony had the pilot connected via his harness's lifting V-ring, and they were both hoisted up 70 feet and secured into the helicopter cabin. The pilots punched in the coordinates for Balboa Naval Hospital and quickly navigated there in the worsening weather. McDonel provided the trembling F/A-18 pilot blankets and water, plus offered him a Subway sandwich that he brought for his own lunch. The overhead P-3 Orion aircrew coordinated an ambulance to be ready at Balboa Hospital to transfer the pilot for medical evaluation. The HS-10 Calumet 621 aircrew landed on the hospital's helicopter pad, transferred the pilot to the ambulance, and then headed back to NASNI to debrief. Earlier that morning when they launched, they had no idea what they were in for. It was just another day of training in the fleet replacement squadron (FRS), for two new students. Then the unthinkable occurred for a USMC pilot and without hesitation the SAR asset promptly responded with professionalism and precision to save a life.

Aviation rescue swimmers will be assigned and must successfully complete their applicable Class "A" School of one of two rates: AWS or AWR. This is different than when I served as we were all AWs (Anti-Submarine Warfare Operator, which later changed to Aviation Warfare Systems Operator and is currently Naval Aircrewman). The current AWS (Naval Aircrewman Helicopter) rate are members of helicopter-integrated tactical crews that mainly perform SAR operations. AWS's are also responsible for aircrew operations administration, flight and ground training, movement of cargo internally and externally, medevac missions, passenger transport assignments, aerial gunnery operations (M-240, GAU-21, M4, M11), small arms handling, naval special warfare (NSW) insertion and extraction, vertical replenishment (VERTREP), and night vision device (NVD) operations. The AWR rate (Naval Aircrewman Tactical Helicopter) operate tactical systems to detect, identify, and engage submarines. They may operate unmanned aircraft systems (UAS) and their payload. They may also man technical support centers at shore installations and tactical mobile units in expeditionary MTOC units (Mobile Tactical Operations Center). AWRs serve as primary mission specializations including HA/DR (humanitarian assistance / disaster relief), HVBSS (helicopter visit, board, search, and seizure), CSAR (combat search and rescue), SAR (search and rescue) assist, SOF support (special operations forces), CAS (close air support), and SUW (surface warfare).

The final phase of the almost two years of training is at the FRS, where candidates are trained to operate in the mission-specific type of helicopter they will be assigned to. Later in their career as aircraft are updated or replaced, aircrew and pilots will return to the FRS to get qualified in the next generation helicopter. It's important for the pilots and enlisted aircrew to learn the material and fly together, both in a simulator and aircraft, since they will be working tightly with one another in their assigned duty stations. The pilots and aircrew are separated in their classroom training except when there's overlap. The aircrew learn the basics of the helicopter and are continually studying and tested on the Naval Air Training and Operating Procedures Standardization (NATOPS). Though they all receive common training on certain SAR procedures, they also require specialized training relevant to the specific aircraft they operate in. They learn the basic essential technology, flight, and safety procedures and then focus on their specialties. These include all the fancy acronyms listed above conducted in classrooms, hydraulic operated simulators, and multiple helicopter day and night flights. During the helicopter training flights, there's always a pilot instructor paired with the pilot trainee and an aircrew instructor paired with the aircrew student. The aircrew / aviation rescue swimmer student typically brings their SAR swimmer's gear on each flight in case they are diverted away from training for a SAR mission. There are also cases when the aircrews aren't SAR capable for equipment reasons or not having their designation letters signed earlier in the training program, but beyond that they tend to fly ready to respond to any local SAR needs.

After months of FRS training, the students who successfully complete the course material, evaluated simulators, and flights will achieve the confidence to fly actual mission-sets in the fleet. However, training never stops since the pilots and aircrew must maintain their qualifications in a myriad of categories. Upon completion of training, aviation rescue swimmers finally earn their coveted gold Navy aircrew wings. It's worth noting that aviation rescue swimmers are one of the only special operations groups that doesn't currently have an approved uniform insignia. Navy aircrew wings are not representative of aviation rescue swimmers but are earned for completing the respective aircraft platform personal qualification standards. There have been conversations for decades about approving the uniform insignia described in chapter four, but as of the writing of this book it remains in limbo.

Hazing has been a military tradition since the beginning of time. Congress officially outlawed hazing in 1874, but the challenge has been properly defining the term *hazing*, whether it's getting punched in the arm when military personnel earn a promotion, to drudging through the

"Polliwog" gauntlet during shellback initiation on the ship as sailors cross the equator for the first time, to my experience earning my wings.

There were only two aircrew / aviation rescue swimmers earning aircrew wings in my 1994 FRS class. We went down to the NAS North Island beach and had a "winging" party. We were both handed overflowing pitchers of cheap beer with our bright, shiny new wings dropped in them. We were then instructed to pound the pitcher without stopping until we retrieved our wings in our mouths. If you vomit during the process, then you better vomit into the pitcher and then drink that as well. A line of veteran aircrew then formed in the hot sandy beach, as one removed the wing's protective backings and punched them directly into each of our chests. One by one the aircrew beat the bent and bloody wings into our chest with a congratulatory handshake following each pounding. I'm not promoting hazing by any means, but I'm personally very proud of those two scars that remain in my chest to this day. I do understand that hazing can quickly get out of hand and has a history of anger, violence, prejudice, harassment, and unfortunate death. Just like anything in life, you must have boundaries and enforce those boundaries to protect others.

AWS and AWR aviation rescue swimmers then receive their orders to one of a few dozen helicopter sea combat or maritime strike squadrons. The squadrons are attached to an air wing, which determines the helicopter platform and ship they'll deploy with. It's at this point that all the training comes together in practical application. The aircrew fly daily mission-sets and qualification flights to ensure they remain in a combat-ready status. Squadrons regularly deploy with Navy warship battlegroups to provide support for the air wing and other military and humanitarian operations, so the aircrew are always ready to leave for weeks or months if the need arises.

In the mid 1990s, halfway through my six-year enlistment, I was in the San Diego Bay conducting helicopter SAR jump qualifications. To maintain our jump qualifications, we were periodically required to jump six times from the helicopter during the day and get "tea bagged" twice at night. That meant we'd get hoisted up 70 feet to the cabin then hoisted back down, then once again to ensure we checked the nighttime requalification box. All the aviation rescue swimmers will typically get qualified at once, which also allowed the pilots to either get the flight qualifications or if they were lucky, jump with us for the experience. We would brief as a team and then split the swimmers up to fly some out to start the jumps while the rest caught a ride on a Navy motor whaleboat.

During this mild summer's evolution, we had some Annapolis students fly out from the Naval Academy to work an internship with our squadron. The U.S. Naval Academy was established in 1845. Its acceptance rate is highly competitive at 9 percent, making it one of the most prestigious U.S. institutions. Midshipmen are highly educated for service in the officer corps of the U.S. Navy and U.S. Marine Corps. During the summer, students travel to different active-duty stations to experience the Navy or Marine Corps in action as they further develop their curriculum for their future officer program. The midshipmen spent a few weeks with HS-2, interviewing with the different shops: maintenance, quality assurance, ordnance, parachute rigger, aircrew, and pilot. They were also included in our flight briefs and brought onboard as passengers for various training mission-sets. That year they were also invited to jump with us during our qualification SAR jumps to get a unique opportunity that most can't imagine experiencing. All candidates were in their early twenties, and they were definitely wide-eyed and amped with anxiety for the experience.

Everybody attended the general flight and rescue training brief to understand the timeline, plan, and safety procedures. We then briefed the midshipmen specifically on what to expect from a rescue swimmer perspective. We were very clear to let them know that it may look cool and exciting from a distance or on television, but it is extremely chaotic once you're in the center of the 60-knot rotor wash and screaming high frequency of the HH-60H twin turboshaft engine and main rotors. Not to mention the cold temps of the water, the potential wildlife below, and possible pollutants, being so near Tijuana, Mexico, the electrocution risk of the static discharge from the hoist cable, the complexity and discomfort of being attached to the rescue strop, the fact that the rescue swimmers would be submerging to do our procedural checks on them, the pain of being hoisted at high speed of 215 feet per minute up 70 feet until the variable speed shock loads us near the top, and the difficulty of maneuvering in a small wildly shaking cabin where it is impossible to hear. That was just to get hoisted up! Then we described how to communicate using hand signals in the cabin of the helicopter and how to get prepped in the door to jump as the helicopter drops from hundreds of feet to an altitude of 10 feet and 10 knots. We discussed the proper technique for jumping from a helicopter to land properly, rather than flat on their back or face planting. And then what to do when in the water while the helicopter continues jumping the remaining swimmers and passengers before it comes back around for extraction. The most important piece of advice was to stay calm and let us do all the work. Oh, and have fun!

A boatswain's mate second-class petty officer chugged the old, gray motor whaleboat to a pier near the helicopter squadron's two massive hemisphere hangars built in World War II. We were dressed in our full wetsuits and rescue gear, while the midshipmen were dressed in borrowed wetsuits and emergency flotation. We piled into the boat with all our gear and some snacks. A couple guys brought fishing gear to pass the time while they waited for their turn to jump. We were all reminded and warned to not pull what we did during our previous SAR jumps. We thought it would be funny to carry some of the fish we caught up to the helicopter and stash them in the pilot's flight bags attached behind their seats. And we were right, it was totally funny! We were cracking up as we jumped from the HH-60H and completed our covert operation. However, the pilots didn't find all the fish buried in their bags when they landed later that evening. And since it was a Friday evening, the dead fish stayed in their bags for the entire weekend. On Monday, the guilty rescue swimmers, including me, were put to work cleaning the aviation survival equipment or parachute rigger shop since it reeked like fish. Totally worth it!

The boatswain's mate steered the haze gray boat underneath the 200-foot-tall Coronado Bridge and took position in the San Diego Bay between the Naval Amphibious Base Coronado, where Basic Underwater Demolition / SEAL training is held, and Naval Base San Diego (32nd Street). As we sat idle, humoring nervous chatter from the midshipmen, two HH-60Hs broke the silence and came thumping in from the south.

One immediately dropped in and conducted a SAR pattern, while the other circled the bay. The crew chief slid open the cabin door and tossed out an armed MK-25 smoke. Upon water submersion, the seawater enters the smoke's canister activating the battery, which initiates an electronic squib that ignites the pyrotechnic composition. The plume of smoke provided the pilots a visual locator of the survivors as well as wind conditions. The crew chief called out the distance and direction as the pilots banked the helicopter into a wide circle to fly into the wind for a more stable approach and hover.

You could consistently tell across the faces of the midshipmen that things were about to get real. Seeing the Navy helicopter flying the SAR pattern was their first reality check that they were actually going through with this crazy exercise. The boatswain's mate staggered the midshipmen with an escort rescue swimmer in a long line in the bay, about 50 yards apart. This is to provide enough space to perform a training rescue outside of the rotor wash. There's nothing quite like an abrupt submersion in cold water to increase your heart rate. The initial shock causes your body to react, closing blood vessels in your skin making it hard for blood to flow, which then forces your heart to work harder and increases blood pressure. There are many benefits of this cold plunge process, which is why ice baths have become so popular. My guy and I jumped from the side of the boat into the cool water. He yelled out an excited word of profanity as I immediately locked into focus like flipping on a light switch. Our wetsuits were saturated in the water and activated their warmth, plus provided some buoyancy. One final time, I made eye contact with my midshipman survivor and explained the procedures, so it was fresh in his head. He acknowledged and seemed OK with the plan as he frequently looked past me to see the helicopter hoisting another group in the distance. Once they were securely onboard, the HH-60H departed and flew a lap around the bay and then approached in a 70-foot hover toward our position for extraction.

The thing that's hard to accurately describe, which I've mentioned a few times throughout this book, is the ear-piercing engine sound, amount of force, and rotor wash that the helicopter kicks up as they approach for extraction. It's equivalent to someone throwing thousands of tiny rocks at your face with the deafening scream of standing near a speaker at a rock concert, all while treading water and struggling to breathe. (Normal conversations are at 60 decibels, alarm clocks are at 80 decibels, and anything over 85 decibels for extended periods will result in permanent hearing loss. Concerts are rated at 110 decibels, gunshots at 140 decibels, and aircraft carrier flight deck and U.S. Navy helicopters sustain a level exceeding 150+ decibels. Hearing protection is required, but even with protection, the personnel will experience permanent hearing loss.) It's an intense work environment and very difficult to remain calm and focused for the untrained, which is precisely why aviation rescue swimmers repeatedly operate in these conditions during training to normalize this work setting. Over time, they feel relaxed and oddly seek the comfort of the chaos.

* * *

In my experience as a counselor working with all types of people, I have seen where those who have grown up in chaos and/or abuse might seek out people and/or jobs that continue the pattern of chaos in their life. There's actually a newer term for this called "chaos addiction," which is the thrill of being in or causing chaos to create adrenaline. By growing up in unpredictability, people are conditioned to live in chaos and then end up seeking it. Living in calm is uncomfortable for them.

There may be a higher rate of these individuals who seek more high-risk jobs like the military or first responders. For these individuals, channeling this into crisis situations to do good and help others is a way to use the chaos of their past to function in the chaos of their present. This allows them to make life-changing interventions for future good. I feel the key is to understand your history and motives and to keep them in a healthy state. Then use support and/or therapy to process through all the chaos so it's not coming out at home or being dealt with through other unhealthy high-risk coping outlets like alcohol, drugs, affairs, and abuse. This is a great example of resilience: having a chaotic childhood but using that experience to do good, ultimately finding the calm in the chaos.

* * *

As the helicopter approached, I looked up to gain visual contact and acknowledgment from the crew chief operating the hoist. I gave a thumbs up and yelled at the survivor to follow me and to do exactly what I say. The helicopter flared nose up and the main rotors WHAP! WHAP! pounded air down onto the water, causing an avalanche of rotor wash. I yelled to swim! The survivor's eyes were about as wide as they could be without popping out of their sockets, but he gave it an effort and attempted to follow me. It's like swimming into the eye of a hurricane. Everything in your mind is telling you to swim away from the dangerous rotor wash, but you must fight that urge to flee and power through to reach safety.

We both kicked hard, and I made sure to keep a close distance with the Academy student to ensure he didn't get left behind. He had wanted to try swimming into the rotor wash rather than have me tow him, which is what I would do in a real rescue. But since he was wanting an authentic experience, I was willing to allow it. We eventually broke through the intense diameter of pain and into the calmer center section, which is a small circle directly under the hovering helicopter. This area acts as a shadow of the main rotors, not allowing as intense of downward pressure, which creates a sort of calm. And by calm, I mean calm in comparison since it's still chaotic by anyone's definition.

We treaded water there in the middle area of the rotor wash as the crew chief lowered the hook and rescue strop. I yelled commands and used hand gestures to communicate what to expect next. I needed the midshipman to be aware and stay clear of the lowering hook, as it would be charged with dangerous amounts of static electricity—15,000 to 20,000 volts! The main rotors, passing through particles of dust or water in the air, build electrostatic induction and a triboelectric charge. The total voltage can range from a minor shock to lethal. All I know is I failed one of my Aviation Rescue Swimmer School multi-rescue scenarios because I didn't allow the hook to ground out in the water first before grabbing it. Even though training was conducted inside the Rescue Swimmer training facility using a simulated helicopter tower with no real risk of being shocked, we still had to train as if it were a live wire. The ARSS instructors didn't say a word. They let me get hoisted up with my final victim, complete my advanced first aid, and then told me I failed because I killed myself and my survivor. I learned my lesson and now was diligent about allowing the hook to submerge before reaching up and getting electrocuted in real life.

The bright orange rescue strop attached to the rescue hook caught the heavy wind from the rotor wash and whipped toward our direction, dragging across the water. I made sure we were both clear as the weight of the hook slowed and stopped the strop, grounding it in the bay. Towing my student survivor in a cross-chest carry, I swam toward the floating strop. That's when the midshipman started to panic. His head and shoulders were swiveling back and forth as his fight or flight was kicking in. He was breathing heavy and thrashing his arms around, making his head go under water, which wasn't helping his panicked attitude. I yelled for him to relax, slow his breathing, and let me do the work. But in that state, he couldn't think rationally. With wide eyes and an open mouth being forcefully filled with seawater from the rotors, he lost it. I knew I needed to immediately resolve this conflict by getting him attached to the strop and clear of the water.

I locked him in a controlled cross-chest carry and kicked toward the strop. Using my left hand, I tightly brought him in closer to my body and quickly released my right hand to grab the floating rescue strop. As soon as I released my right hand, the midshipman went active on me, thrashing to get away. An active survivor is when a panicked or aggressive victim sees you as flotation and tries to latch on to stay afloat. When I detected an attack, I instinctively went to life-saving training. I simultaneously grasped a pressure point on the midshipman's left elbow and another under his jaw, applying pressure as I spun him around back into a controlled cross-chest carry. Within a second, he was again locked in my control. The unnerved student was temporarily submissive as I towed him back to the strop. I thought to myself, I wonder if any of the other rescue swimmers were experiencing similar panicked victims?

We approached the orange hoisting instrument and I communicated at the top of my lungs that I needed to release my arm lock to reach the rescue strop. Then as soon as I loosened my grip and grabbed for the strop, he once again attempted to spin out of my hold. I then went a little off script. With my right hand, I grabbed the rescue hook, pulled the student toward me with my left arm, and smacked the disobedient midshipman in the forehead with a couple pounds of forged, stainless steel. For a brief moment he was stunned, which I took full advantage of and wrapped the rescue strop securely around the victim's back and swiftly connected the other end to the rescue hook. I then connected the safety strap around his chest at which time he was locked in. I connected my own harness-lifting V-ring to the rescue hook and forcefully kicked backward with my fins, pulling the survivor and hoist cable taught. While doing this, I extended my right arm and gave a thumbs up, indicating to the crew chief that he was cleared to hoist. The whole incident took a matter of only a few seconds. We train for the things that are within our control, but we must be prepared to respond to things that are outside our control.

* * *

"God grant me the serenity to accept the things I cannot change, the courage to change the things I can, and the wisdom to know the difference." This is referred to as the Serenity Prayer and was authored by American theologian Reinhold Niebuhr (1892–1971). It's often memorized and recited for those struggling with addiction, as used in most all twelve-step recovery support groups, such as Alcoholics Anonymous (AA). But it's a powerful prayer for anyone.

Many people struggle with the need to control their environment. For those who come from a history of child abuse, trauma, and/or have PTSD, the need can be even greater. By trying to control their environment they might be unconsciously trying to prevent further trauma, trying to protect themselves from the things they once experienced. They must focus on the things within their control: thoughts, beliefs, feelings, and actions. Give the rest to God and pray the Serenity Prayer.

* * *

As the crew chief pressed up on the hoist hover trim, we both dragged across the water until directly under the hovering HH-60H. We were then both lifted from the water, and I tightly wrapped my legs around the survivor to keep him locked to my body for the duration of our trip up 70 feet to the helicopter cabin. I shouted on the way up to let him know that he was safe and joked that I forgave him for attacking me. Eyes dilated and still in shock and a line of blood running down his face originating from his forehead, he transitioned past his moment of panic and thanked me profusely for whatever I just did to get him through that. We reached the helicopter's cabin where the crew chief looked at me with a smirk as he lowered us inside. I helped us both detach from the rescue hook and secured our gunner's belts to ensure we didn't accidently fall out of the moving helicopter. The pilots then moved the hovering aircraft forward to hoist up the remaining rescue swimmers and student survivors. Again, I wondered if any of the other Annapolis students were as entertaining as mine.

When all were safely onboard, we flew a wide circle pattern around the San Diego Bay. From the looks in the students' eyes, they seemed pretty amped up and nervous for the next evolution. They wouldn't have much time to contemplate as the pilots pulled the weighted-down helicopter back into the wind and quickly dropped to a safe altitude and speed to jump. My midshipman and I were the first to go as we sat in the cabin door with our legs hanging out. The rotor wash kicked up spray, engulfing the view as we approached 10 feet and 10 knots. The crew chief tapped my shoulder three times, and I exited the aircraft. I gave an OK hand signal from the water and watched as the midshipman plunged into the bay. I quickly swam toward him to make sure he was good as the HH-60H continued dropping swimmers and students down a straight line. The motor whaleboat drifted in to gather the Annapolis guys, and we rescue swimmers remained in the water to continue with our qualification SAR jumps.

Later that evening at the squadron hangar, I met up with the student I was partnered with. The entire experience was a lot to process, but he was grateful and would never forget it. He admitted that he underestimated the chaos that occurs in the water under a hovering helicopter. It was a good example of how quickly a perceived safe situation can turn to panic even in a controlled training evolution. In that scenario, the student wasn't in any real risk since he had flotation and a "professional lifeguard" at his side. I was able to distract him with a hook to the face to get him past his moment of frenzy. Once hoisted from the water he was able to calm down and had an experience of a lifetime jumping from the Navy helicopter.

There are so many areas in our lives that we can get caught off guard during our normal routine. Our safe world is thrown off balance and we tend to panic. During those moments, we lack control and react in a way that can be life threatening to our relationships or well-being. Sometimes all we need is a metaphorical smack in the head by a rescue hook to get distracted from the perceived harm. This forces us to rise high above the danger to focus on the bigger picture.

* * *

A great tool to help increase focus is to practice mindfulness. It's been shown that doing this on a regular basis reduces stress hormones. It can also help you get to a place of peace when fight, flight, or freeze is triggered. You can easily do this by sitting in a quiet environment and closing your eyes. Focus on your breathing and the sounds and sensations around you. This includes focusing on your five senses. What do you hear? What do you taste? What do you smell? What do you feel? And finally, open your eyes. What do you see? By focusing on the present moment, you rewire the brain and gain more control over your focus and attention. You can then use this tool to calm your nervous system when it is flooded or whenever you need to reset yourself.

* * *

On April 24, 1988, the USS *Bonefish* caught fire 160 miles off the coast of Florida. The Barbel-class submarine, commissioned in 1959, was participating in anti-submarine training exercises with the USS *John F. Kennedy* aircraft carrier, its anti-submarine warfare helicopters, and the USS *Carr*. The diesel submarine had just completed recharging its 506 lead-acid batteries and was preparing to dive when one of their engineers noticed one of the batteries had grounded. At the same time, due to a leaky seal in the *Bonefish*'s trash disposal unit, saltwater gradually began to build up. As the submarine dove, the angle of the pitch allowed the saltwater to corrode the exposed battery cables. This hazardous combination caused overheating in the forward battery, which led to red glowing power cables arcing across the battery bus and catching fire. (Diesel submarine engines are combustion engines, which require a steady stream of oxygen to run. While submerged, diesel submarines run on electric motors, which are powered by batteries that charge while the diesel engine is running. Due to the lack of oxygen in the ocean, diesel submarines need to surface or extend a snorkel above the water to replenish the required air, where a nuclear submarine can remain submerged almost indefinitely.)

The crew unsuccessfully attempted to saturate the flames with their CO_2 fire extinguishers and then were ordered by the skipper to seal the hatch, trying to choke out the flames. That's when a massive explosion set off, sending a fireball through the compartment. The enormous jolt sent sailors flying across the submarine, injuring several. The *Bonefish* was on fire both inside and outside, while it was still submerged. The skipper ordered an emergency surface as the crew was blinded and asphyxiated by the dense, black smoke of the fire. The crew donned emergency air breather face masks and made their way to any available hatch to abandon ship. Sadly, three sailors succumbed to injuries and suffocation from the smoke.

The USS *Carr* aggressively steamed toward the plume of smoke and began coordinating rescue efforts. They deployed life rafts, as there were multiple sailors thrashing about in the oil-infested water. The USS *John F. Kennedy* diverted a plane guard helicopter from Helicopter Anti-Submarine Squadron 7 (HS-7) to assist in the rescue. The SH-3H Sea King helicopter arrived on scene and immediately began hoisting survivors. Twenty-one-year-old aviation rescue swimmer AW3 Larry Grossman was on Alert 30 on the aircraft carrier when he heard, "The BONEFISH is on fire! The BONEFISH is on fire!" screaming through the ready room intercom. He and his aircrew, Lieutenant Commander Andy Krug, Lieutenant Junior Grade Wes Reel, and Crew Chief AW1 Chris Carnes, had already briefed and hustled to the Alert helicopter. They patiently waited on the port side catwalk since their aircraft was stowed on the starboard side of the flight deck. They had to wait for all the fixed-wing aircraft to land before crews could tow the rescue helicopter. In the far distance, they watched smoke clouding the sky from the surfaced submarine and grew with anxiety, eager to help.

The pilots conducted their pre-flight checks as they spun up the main rotors. Within a few minutes they were airborne and on station to assist the other HS-7 SAR asset. Grossman recalls the complete chaos unfolding in front of his eyes as they searched for survivors. It looked like a traumatic war scene from a movie with the submarine ablaze with black smoke clouds ferociously pumping into the sky. Screaming sailors thrashed in the high seas, trying to climb into rafts with the ocean surface saturated with fuel. Everyone was in shock and panicked. It was hard to prioritize who to rescue first. The pilots just picked a raft full of sailors and pulled the Sea King into the wind and prepared to deploy the young rescue swimmer.

As daunting as the situation appeared, Grossman remembers the best feeling in the world were those three taps on his shoulder from his crew chief. This was the indication that it was safe for him to jump from the hovering helicopter. The rescue swimmer checked for debris and then pushed off from his sitting position in the SH-3H cabin and free-fell to the Atlantic Ocean. As Grossman entered the 77-degree, shark-infested water, his adrenaline was jacked up, but he compartmentalized all fear and focused on the mission of saving lives. This was the moment he trained for, and he was about to be put to the ultimate test.

Grossman swam through the fierce rotor wash and then 50 yards to the nearest drifting life raft. Nightmarish screams and moans echoed from the bright yellow raft filled with terrified and burnt survivors. The HS-7 helicopter kept a safe distance to allow the rescue swimmer to do his work, but to also avoid capsizing the raft from the heavy rotor wash. Grossman communicated loudly that he was there to help but needed to prioritize the most critically wounded first. He spent a few moments assessing the twenty-six submarine crewmembers in the raft. They were all covered in black oil and coughing from extreme smoke inhalation.

Grossman identified a critically burnt victim and demanded he roll out of the raft and into the water. The terrified victim refused. The rescue swimmer responded by using a technique perfected in training. In one fluid motion, he lifted himself up on the raft's side flotation, grabbed hold of the sailor's dungaree uniform, and pulled him backward into the ocean. The shocked survivor hit the water before he realized what was happening. Unable to swim, he thrashed around in a panicked state and then sank deep into the darkness. Grossman immediately dove down 5 feet, kicking his fins for speed, and gained control of the drowning victim in a cross-chest carry. He towed him back to the ocean surface where they both took a deep breath of air, and the survivor coughed out inhaled seawater. The panicked victim then spun his oily body around and went active, desperately trying to use the rescue swimmer as flotation. His flailing arms knocked off Grossman's mask, which sunk to the bottom of the Atlantic. Grossman remembers using a front head hold release life-saving procedure to gain control of the survivor. Performing the tactical move on the surface, he used a series of pressure points and movements to turn the panicked survivor around to regain control. The submarine sailor finally calmed down long enough for Grossman to give him his personal SAR-1 flotation. At that point, he was able to tow the sailor toward the hovering SH-3H. The survivor remained submissive as they entered the rotor wash, which was blinding for both as Grossman was now without a mask. He found the rescue strop and secured his victim and then gave a hand signal to the crew chief to hoist him up. As Grossman released the saved victim to be hoisted, he turned to swim back toward the chaos and remembers hearing the sailor shout, "Thank you for saving my life!"

At this point, the life raft was floating farther away from Grossman. He waved his hands to get the crew chief's attention and then signaled for them to short-haul him closer. Carnes acknowledged his hand signals and lowered the hoist, and Grossman connected his harness-lifting V-ring to the rescue hook. The crew chief lifted the rescue swimmer clear of the ocean, and the pilots hauled him closer to the life raft, where he was lowered back into the water. Grossman continued to deliver survivors to the helicopter, one by one. With eight sailors safely hoisted to the SH-3H, they needed to return to the aircraft carrier to offload the survivors and refuel. Grossman remained in the water.

Evening was quickly setting in, and there were still several survivors who needed to be rescued. All six HS-7 helicopters were rotating in the action with three rescue swimmers in the water. They hoisted the exhausted victims up to the hovering helicopters until they were full of passengers, and then they would shuttle them back to the USS *John*

F. Kennedy where they received medical attention. Grossman had been in the water actively saving lives for a few hours, but relentlessly continued plucking sailors from the rafts. He rode the adrenaline rush well beyond its natural dose. He fought the high seas, towing victim after victim into the intense rotor wash of the hovering SH-3H. With limited vision, he used his sense of touch to ensure each survivor was securely strapped into the rescue strop to be hoisted up 40 feet to safety. Without his mask or emergency SAR-1 flotation, the salty sea, dense smoke, and heavy fuel in the water severely burned his eyes. The intense pain from his virtually swollen shut eyes was almost unbearable, but he never considered stopping. He felt tired, but the lives of fellow shipmates depended on his ability to continue his rescue efforts. Sacrificing his own safety, he continued saving lives.

Three exhausting hours passed with Grossman still operating in the open ocean. Finally, he towed his last survivor into the hurricane force rotor wash to the awaiting rescue strop. The rescue swimmer had swum well over a thousand yards in the harsh ocean conditions and was credited with rescuing nineteen sailors. Completely fatigued, he hoisted up with the final survivor 40 feet to the hovering SH-3H. As he cleared the water, he recalls spinning and was too weak to self-stop. He then passed out. The crew chief hoisted the limp rescue swimmer and the final survivor to the cabin door. He carefully lowered them both onboard and strapped a gunner's belt around the submarine sailor. He then placed a SAR-1 and and helmet on Grossman before strapping him into the rescue litter. Grossman was in and out of consciousness. The only thing he recalls hearing through his helmet's ICS were the pilot's radio calls back to the carrier, "Clear the flight deck! We have a rescue swimmer down!"

The USS *John F. Kennedy*'s mess deck was cleared for mass casualty and triage. After the HS-7 aircraft was safely chocked and chained to the flight deck, Grossman was carried on the rescue litter to the ship's medical ward. He couldn't stand, walk, or see. His body was in a physical state of shock as he was treated for exhaustion, dehydration, minor burns, and severe eye irritation. Eighty-nine sailors were rescued from the USS *Bonefish*, with three casualties. Larry Grossman performed one of the most heroic and selfless rescues in U.S. Navy aviation rescue swimmer history, making him a legend in the SAR community. He was awarded the Navy and Marine Corps Commendation Medal for heroism, which was later upgraded to the Navy and Marine Corps Medal by the president, the highest non-combat decoration awarded for heroism by the U.S. Department of the Navy.

Larry Grossman remained calm in the chaos due to his extensive training and "never say can't" attitude he says he learned from his father. As his helicopter descended to a safe altitude to jump, he sat in the doorway assessing the chaotic scenario he was about to engage. The situation would be overwhelming for anyone, but he was trained to identify and prioritize the survivors based on their critical needs. In Aviation Rescue Swimmer School "multi" training, you typically have two pilot survivors and one free floater. Grossman took this training to the extreme by rescuing nineteen free floaters in a three-hour period, at night, and without his essential safety equipment. Nobody truly knows how they'll respond when put to the test, but it's in those moments that truly define your character. Grossman and his HS-7 peer rescue swimmers put the lives of the desperate USS *Bonefish* sailors above their own. Because of their selfless choice to focus and serve others, eighty-nine sailors were able to see their families again.

Aviation rescue swimmers go through some of the most grueling and unforgiving training and mission-sets in the U.S. military. It takes years of perfecting various techniques, maintaining strong physical fitness, being proficient with specialized equipment, gaining new knowledge, and confident decision-making to become an efficient leader. It's that unrelenting focus that allows us to not only survive the impossible but to also save lives in the process. Although only a small number will earn the elite status of a U.S. Navy aviation rescue swimmer, there are valuable lessons everyone can glean from our attitude, training, and experiences. Lessons that can help us all be calm in the chaos!

Bibliography

"160th Special Operations Aviation Regiment (Airborne)." Wikipedia, October 6, 2022, en.wikipedia.org/wiki/160th_Special_Operations_Aviation_Regiment_(Airborne).

"2015.0053.3—17 Color and B&W Photos of Apollo 13 Recovery. Photos Not Catalogued Individually." UDT-Seal Museum Association, navysealmuseum.pastperfectonline.com/photo/23380BD6-9486-435B-A9BF-443578524015.

A., Nicholas. "Navy Aircrewman: Career Details." Operation Military Kids, June 27, 2019, https://www.operationmilitarykids.org/navy-aircrewman-career-details/.

Abramson, Ashley. "Cultivating Empathy." American Psychological Association, 2021, www.apa.org/monitor/2021/11/feature-cultivating-empathy.

"APA PsycNet." American Psychological Association Psycnet, psycnet.apa.org/record/2000-15337-006.

"Are Diesel-Powered Submarines Better Than America's Leading Nuclear Fleet?" *Observer*, October 18, 2019, observer.com/2019/10/diesel-powered-submarines-vs-american-nuclear-fleet/.

Bauers, Sandy. "Rescue Swimmer Battles Darkness, High Waves to Save 2 Pilots." *McClatchy DC*, May 24, 2007, amp.mcclatchydc.com/latest-news/article24436465.html.

"The Benefit of Spirituality on Our Well-Being." *Psychology Today*, www.psychologytoday.com/us/blog/hope-resilience/202101/the-benefit-spirituality-our-well-being.

"Bill Rutledge, Det 8—War Story." U.S. Navy Seawolves, seawolf.org/war-story/bill-rutledge-det-8-war-story/.

"Blink: The Power of Thinking without Thinking." Wikipedia, January 17, 2023, en.wikipedia.org/wiki/Blink:_The_Power_of_Thinking_Without_Thinking#:~:-text=Blink%3A%20The%20Power%20of%20Thinking%20Without%20Thinking%20(2005)%20is.

Casteel, Chris. "Faith in God, Family Cited by Ex-POW." *Oklahoman*, March 15, 1991, www.oklahoman.com/story/news/1991/03/15/faith-in-god-family-cited-by-ex-pow/62533952007/.

Cherry, Kendra. "How Resilience Helps with the Coping of Crisis." *Verywell Mind*, October 17, 2022, www.verywellmind.com/what-is-resilience-2795059.

"The Code." YouTube, www.youtube.com/watch?v=3zUImnnjCtI.

Correll, Diana Stancy. "'I Just Did What I Was Trained to Do': New Book Chronicles Life of One of Last Two Arizona Survivors." *Navy Times*, February 16, 2021, www

.navytimes.com/news/your-navy/2021/02/16/i-just-did-what-i-was-trained-to-do-new-book-chronicles-life-of-one-of-last-two-arizona-survivors/.

"Decibel Chart of Common Sounds | DB Comparing Decibel Levels." Decibel Meter App | Best Digital Sound Level Meter for Your Smartphone, December 20, 2021, decibelpro.app/blog/decibel-chart-of-common-sound-sources/.

Definition of *trauma*. Merriam-Webster.com, 2019, www.merriam-webster.com/dictionary/trauma.

Dickey, Christopher. "Military Justice: Saddam's Crimes." *Newsweek*, February 16, 2003, www.newsweek.com/military-justice-saddams-crimes-140519.

"Fear." *Psychology Today*, 2019, www.psychologytoday.com/us/basics/fear.

Foster, Gary Wayne. "The Hanoi March." Hellgate Press, 2022, www.hellgatepress.com/wp-content/uploads/2022/03/Pages-from-THM.pdf.

Garamone, Jim. "Remembering Hurricane Katrina a Decade Later." U.S. Department of Defense, August 25, 2015, www.defense.gov/News/News-Stories/Article/Article/615149/remembering-hurricane-katrina-a-decade-later/#:~:text=More%20than%2060%2C000%20members%20of.

"GI Film Festival San Diego | Scramble the Seawolves." KPBS, video.kpbs.org/video/scramble-the-seawolves-yacuzi/.

Goleman, Daniel. *Emotional Intelligence*. New York: Bantam Books, 1995.

Greenstein, Luna. "The Mental Health Benefits of Religion & Spirituality." National Alliance on Mental Illness, December 21, 2016, www.nami.org/Blogs/NAMI-Blog/December-2016/The-Mental-Health-Benefits-of-Religion-Spiritual.

"Gulf War POWs Tell of Saddam's Wrath." ABC News, abcnews.go.com/amp/2020/story?id=123740&page=1.

"Helicopter Combat Support Squadron Seven." Wikipedia, March 28, 2022, en.wikipedia.org/wiki/Helicopter_Combat_Support_Squadron_Seven.

"HSC-84." Wikipedia, November 9, 2022, en.wikipedia.org/wiki/HSC-84.

"Hurricane Katrina Helicopters." Helis.com, 2005, www.helis.com/featured/katrina.php.

Jennewein, Chris. "Helicopters Make 1,500 Water Drops in Battle to Save USS Bonhomme Richard." *Times of San Diego*, July 15, 2020, timesofsandiego.com/military/2020/07/15/helicopters-make-1500-water-drops-in-battle-to-save-uss-bonhomme-richard/.

Jones, Rebel. "35 Viktor Frankl Quotes on Challenges, Suffering & Finding Purpose." *Happier Human*, October 17, 2022, www.happierhuman.com/viktor-frankl-quotes/.

"MILPERSMAN 1220-010." MyNavy HR, August 30, 2021, www.mynavyhr.navy.mil/Portals/55/Reference/MILPERSMAN/1000/1200Classification/1220-010.pdf?ver=wzovGZKvQIzaNIwRG0QkKQ%3d%3d history.nasa.gov/afj/ap13fj/29day6-returnhome.html.

"Mullen Awards HS2 Sailor the Navy/Marine Corps Medal." DVIDS, www.dvidshub.net/news/22969/mullen-awards-hs2-sailor-navy-marine-corps-medal.

"NAS Whidbey Island History." Commander Navy Region Northwest, cnrnw.cnic.navy.mil/Installations/NAS-Whidbey-Island/About/History/.

"Naval Air Station Whidbey Island." Wikipedia, March 25, 2023, en.wikipedia.org/wiki/Naval_Air_Station_Whidbey_Island.

Nees, J. D. "Coronado Eagle and Journal 20 June 1991." California Digital Newspaper Collection, June 20, 1991, cdnc.ucr.edu/cgi-bin/cdnc?a=d&d=CJ19910620.2.76&e=-------en--20--1--txt-txIN--------.

Ortiz, Miguel. "The US Navy Has Its Own Helicopter Squadron Dedicated to Supporting Special-Operations Missions." *Business Insider*, www.businessinsider.com/the-navy-has-a-squadron-dedicated-to-special-operations-support-2021-5.

OuinetAdmin. "US Navy Loses Puma Helicopter." *Defense-Aerospace*, April 10, 2003, www.defense-aerospace.com/us-navy-loses-puma-helicopter/.

"Perspective—Yahoo Search Results." search.yahoo.com/search?p=Perspective+&fr=yfp-hrmob&fr2=p%3Afp%2Cm%3Asb&.tsrc=yfp-hrmob&ei=UTF-8&fp=1&toggle=1&cop=mss.

Pinsky, James. "Dive Motivator: Starting off on the Right Flipper." Free Library, August 1, 2004, www.thefreelibrary.com/Dive+motivator%3A+starting+off+on+the+right+flipper+...-a0117035511.

Price, Alfred. "To War in a Warthog." *Air & Space Forces Magazine*, August 1, 1993, www.airandspaceforces.com/article/0893warthog/.

"The Rime of the Ancient Mariner." Wikipedia, November 15, 2019, en.wikipedia.org/wiki/The_Rime_of_the_Ancient_Mariner.

Roblin, Sebastien. "A Fireball: This U.S. Navy Submarine Was on Fire Underwater." *National Interest*, June 19, 2021, nationalinterest.org/blog/buzz/fireball-us-navy-submarine-was-fire-underwater-188172.

"Rotor Review Fall 1988 #23 by Naval Helicopter Association, Inc." Issuu.com, 1998, issuu.com/rotorrev/docs/23.

"SAR Corpsmen Remembered by Marines, Sailors." Marine Corps Air Station Beaufort, www.beaufort.marines.mil/CommStrat/News/News-View/Article/523706/sar-corpsmen-remembered-by-marines-sailors/.

"Serenity Prayer." Wikipedia, December 20, 2019, en.wikipedia.org/wiki/Serenity_Prayer.

"The Significance of the Serenity Prayer in Recovery Support Groups." *Verywell Mind*, www.verywellmind.com/the-serenity-prayer-62614.

SingleCare Team. "Anxiety Stats in the U.S." *Checkup*, May 6, 2020, www.singlecare.com/blog/news/anxiety-statistics/.

Sterner, Doug. "Ernie Crawford—Recipient." Wall of Valor Project, *Military Times*, valor.militarytimes.com/hero/5748.

Stinson, Adrienne. "Box Breathing: How to Do It, Benefits, and Tips." *Medical News Today*, June 1, 2018, www.medicalnewstoday.com/articles/321805.

Sutton, Jeremy, PhD. "What Is Emotional Awareness? 6 Worksheets to Develop EI." PositivePsychology.com, November 24, 2021, positivepsychology.com/emotional-awareness/.

Swoope, Jan. "Former Vietnam POW to Share Experiences at Town and Tower Breakfast." *Dispatch*, January 29, 2012, cdispatch.com/lifestyles/2012-01-29/former-vietnam-pow-to-share-experiences-at-town-and-tower-breakfast/.

"Tips to Improve Concentration." *Harvard Health*, October 1, 2020, www.health.harvard.edu/mind-and-mood/tips-to-improve-concentration.

"Top 25 Adversity Quotes (of 1000)." A-Z Quotes, www.azquotes.com/quotes/topics/adversity.html.

"Trauma Triad of Death." Wikipedia, November 15, 2021, en.wikipedia.org/wiki/Trauma_triad_of_death.

"UDT's and the Space Flight Programs." National Navy UDT-SEAL Museum, www.navysealmuseum.org/naval-special-warfare/udts-space-flight-programs.

U.S. Navy. "Air Test and Evaluation Squadron (VX) 1." 2023, www.airlant.usff.navy.mil/vx1/.

"USS Bataan and USS Elrod Rescue 282 People in Mediterranean." Defense Media Network, www.defensemedianetwork.com/stories/uss-bataan-and-uss-elrod-rescue-282-people-in-mediterranean/.

"USS *Bonefish* (SS-582)." Wikipedia, May 21, 2022, en.wikipedia.org/wiki/USS_Bonefish_(SS-582).

"USS Ranger (CV 61) History." Uscarriers.net, 2015, www.uscarriers.net/cv61history.htm.

"USS Theodore Roosevelt CVN-71 Nimitz Class Aircraft Carrier US Navy." Seaforces.org, www.seaforces.org/usnships/cvn/CVN-71-USS-Theodore-Roosevelt.htm.

"VX-6." Wikipedia, January 18, 2023, en.wikipedia.org/wiki/VX-6.

"Whidbey SAR Rescues Hiker near Snoqualmie Pass." *Homeport Northwest*, July 23, 2013, homeportnorthwest.wordpress.com/2013/07/22/whidbey-sar-rescues-hiker-near-snoqualmie-pass/.

"Whidbey Team Airlifts Man from Mt. Baker." *Whidbey News-Times*, July 1, 2000, www.whidbeynewstimes.com/news/whidbey-team-airlifts-man-from-mt-baker/?fbclid=IwAR3kHSqeo3scowfbHYtqO8udogJv3FH4y82bwqI0lDB6YoOkDXnJzQUkAk4.

"Why Do You Need to Control Everything? 5 Causes of Controlling Behaviors." *Psych Central*, March 29, 2022, psychcentral.com/blog/why-you-need-to-control-everything#why-you-need-to-control.

Yaede, Jonathan. "Rotor Review Fall 2015 #130 by Naval Helicopter Association, Inc." Issuu.com, 2015, issuu.com/rotorrev/docs/rr130_digital.

Ziezulewicz, Geoff. "This Navy Rescue Swimmer Saved Lives After a Boat Capsized in San Diego." *Navy Times*, July 13, 2022, www.navytimes.com/news/your-navy/2022/07/12/this-navy-rescue-swimmer-saved-lives-after-a-boat-capsized-in-san-diego/.

Acknowledgments

This book has been a culmination of thoughts and ideas I've avoided for a few decades. Although my military experiences were mostly positive, they also included life-impacting trauma and loss, something I and many others have suppressed through the years hoping time will heal. Eventually time doesn't heal, and we're faced with the past and the harsh and scary grip it can have on us. After interviewing over thirty heroes and hearing vivid details of each unique experience, I quickly realized that I was not alone. The conversations between brothers and sisters, spanning generations, were mutually therapeutic. When you are a part of a small, elite group, you can't expect the average person to understand. As a result, we tend to keep a lot of it within. I had to pause a few times over the past years of interviewing and writing to ensure I found help for the issues and trauma I buried years ago. I'm so humbly grateful for the unlimited grace of God and His consistent willingness to carry my burdens, no matter the size or reason.

I was repeatedly told by people, both included and not included in this book, that our conversations immensely helped in just being able to share openly with someone who gets it. And because of it, we now have unconditional trust and lifelong friendships. Having walked it out and continuing to walk it out, I know that there is hope and healing. It's a journey, but worth it to create a healthy life and healthy relationships. I pray that those silently struggling with their own traumatic experiences will seek help. There's so much life to live and you deserve happiness!

First off, thank you to all the past "**SAR OGs**" who paved the way in making the aviation rescue swimmer job, training, and community what it is today. Nothing exists without the blood, sweat, tears, and determination of a few individuals that have an idea and the courage to implement it. And then those courageous folks who commit to following their leadership during times of limited resources and support.

I'm so humbled and grateful to have had a conversation with **Bill Rutledge** shortly before his passing. It is a gift that I will keep with me for the rest of my days. He was as badass as they come, and he and the HA(L)-3 Seawolves created a necessary movement during a time of need that has evolved into today's high-demand Navy CSAR program.

Although not included in the book, I want to thank **Tony Discenso** for everything he taught me during our time served together at HS-2. Anyone who had the pleasure of knowing Tony knows what an amazing leader he was. Although a "ballistic BB," he had a heart of gold and would do anything for the people who worked for him, including never letting me dress as the swimmer on our flights because *he wanted that rescue*! Tony was taken from this earth too early, but his legendary status remains forever in the SAR community!

Thank you for allowing me to spend time with you and to share a small glimpse into your incredible wisdom and experiences **Drew Worth, Taryn Frazelle, Mike Rogers, Whitney Warren, Jesse Hubble, Justin Tate, Charles Raygor, Melissa Dixon, Hector Rodriguez, Jeff Bast, Charlie Harcus, Marty Crews, Joe Sutherland, Bill Gibson, Kevin Gordon, Larry Grossman, Shawn Porter, Tim Hawkins, Scott Chun, Gregory Highfill, Mike Longe, Jay Shropshire, Jeremy Burkart, Cale Foy, Rich Pentony, Michael Bulman, Jonathan Showerman, Ed Majcina, Cory Merritt**, and **Megan Buriak**.

Thank you for being a sounding board and providing valuable feedback and accuracies in my journey of re-exploring the past, while trying to remain relevant with the rapid growth of the present SAR program **Jeff Strickland, Jason Barney, Robert Holloway, Keith Redmond, Jaxson Ingraham, Mike Ousley, Ron Eberly, Sean Frankie "Coco" Coburn, Chris "Val" Valdez, Luis Borges**, and for all those who asked to remain anonymous but willingly provided invaluable conversation, feedback, and draft edits. Also, thank you to the open and honest private **Facebook forum for past and present SAR swimmers**! I know there are countless untold rescues that deserve the spotlight. I wish I could have included them all. I hope this book provides encouragement for others to write and share more stories.

Thank you to my literary agent(s) **Dave Schroeder**, **Brian Mitchell**, and the **WTA Media team** for sticking with this as I explored and developed the idea. It was a long process, but you always believed in what I was trying to accomplish and didn't try to alter the vision of sharing inspiring stories to help others in their personal life struggles.

Thank you **Lyons Press**, **Globe Pequot**, and **Rowman and Littlefield**, for seeing the possibility of this project and turning it into something tangible. Thank you **Eugene Brissie**, **Lynn Zelem**, **Jason Rossi**, **Joshua Rosenberg**, and everyone who tirelessly works behind the scenes to make it all possible!

Thank you, **Chris Razoyk** and the **Rescue Swimmer Shop**, for partnering on collaborative media!

Thank you to bestselling authors **Eric Blehm** and **Jack Carr** for your inspiring work and memorable conversations.

Thank you **Chief Aviation Survival Technician Kevin Cleary** for providing media of the current life-saving techniques taught in the U.S. Coast Guard Rescue Swimmer School.

Thank you **Captain Jim Gillcrist** and the **National Helicopter Association** and ***Rotor Review* magazine** for your invaluable resources and continued support.

Thank you to my children, **Emily** and **Jordan**, for your patience and understanding throughout this long journey. You guys are awesome and always remind me of what life is all about!

And to my amazing wife, **JoAnna**, for your thoughtful understanding and support all these years. I could not ask for a better life partner to walk out this journey together. I'm thankful you finally succumbed to my exhausted begging for your wisdom to be included in this book! I'm so grateful to have your years of education, experience, empathy, and intelligence intertwined in each chapter! The world is truly a better place because of you!

So Others May Live